Redefining Gravity

David Michalets

Self-published on **October 13, 2020**

Table of Contents

Introduction

Redefining Gravity is a comparison of Newton's force of gravity with Einstein's space-time curvature.
The goal is the correct definition of gravity, with the choice of either a physical fundamental force or a mathematical distortion of a desired trajectory of the special observer.

Gravity has two different interpretations, first from Newton and later from Einstein. A clear description of their approaches is followed by the author's proposed mechanism of Newton's force.

Newton had defined the force of gravity but not its mechanism, enabling Einstein to explain a gravitational effect with no force.
With that defined mechanism, Newton's force can return to the status of the correct explanation of gravity. Einstein's space-time had evaded Newton's force with no evidence that a complicated special reference frame behavior was better than a force.
Claims of relativity being confirmed in all cases are refuted.

In the process, gravity is redefined to be a physical force, not a change in path of a special moving observer.
This is an alternate view of this fundamental force in physics.
Isaac Newton defined the force of gravity many years before James Clerk Maxwell and others worked with the fundamental electric and magnetic forces.

This author proposes gravity has a similar mechanism to another fundamental force, the electric force between charges.

Pursuing this fundamental force of gravity between masses included evaluating the assigned values of mass to the electron and proton.

Among the other topics are:
Matter and the atomic model,
Mass defect,
Condensed matter,
Particle pair production,
Kepler's laws of motion,
Black holes, dark matter, precession,
Gravitational lensing and waves.

This is a brief summary of the respective sections:

1) Basics of matter and atoms,
   including mass values of electron and proton,
   particle pair production,
   and velocity limit on mass,
2) Thorough analysis of mass defect,
   through all the elements,
3) Author's atomic model
   by using observations of mass defect,

4) Author's mechanism driving Newton's force of gravity,
   thought to be a mystery.
5) Cases of Isaac Newton's force of gravity.
6) Explanation of Johannes Kepler's laws of planetary motion,
   including a change to the third law.
7) Basics of Albert Einstein's theory of general relativity.
8) Explanation of relativity's space-time.
9) Explanation of the claimed gravitational wave.
10) Explanation of the claimed gravitational lensing,
    also known as gravity bending the path of light.

11) Explanation of the claimed black hole.
12) Explanation of the claimed neutron star.
13) Explanation of gravitational precession.
14) Explanation of dark matter.
15) Conclusions from the above topics.
16) All Internet references in this book can be found in a page in the author's web site, identified here, on the book's last page.

Each page having an external reference has a link to the reference, whether a web page, pdf, or YouTube video. A reader of the book with access to the Internet can read original reference material and view their high resolution images, or watch a video. This book offers excerpts, so reading the references is not required.

One should note the author's first book, Observing Our Universe, had an extensive description of relativity not conforming to its claims. The topic of gravitational waves is relevant to both books. A little of that material is included in this book, so that book is not required for this one.

1 Basics of Light and Matter

Light is in the visible frequency part of the spectrum of radiated energy which is called electromagnetic radiation. This term comes from its mechanism, which is synchronized, perpendicular, electric and magnetic fields which propagate through a vacuum at a velocity measured and called the constant c. This measurement used our defined increment of time, a second, and resulted in the definition of the 1 meter unit of distance.

James Clerk Maxwell explained the propagation mechanism. His equations will be mentioned later.

The explanation of gravity requires analysis to subatomic particles which make up matter, and how electromagnetic radiation is generated by matter.

1.1 Matter

Matter is every object we can see or measure. Each object is radiating, reflecting, or absorbing energy.

Any particle of matter holding an electrical charge is also called plasma. Plasma has unique behaviors compared to matter having no charge, because of the interaction of electric and magnetic fields. Matter consists of atoms and their molecules, as well as any subatomic particles having mass, such as electron, proton, and neutrino.

Excerpt from Wikipedia:

In classical physics and general chemistry, matter is any substance that has mass and takes up space by having volume. All everyday objects that can be touched are ultimately composed of atoms, which are made up of interacting subatomic particles, and in everyday as well as scientific usage, "matter" generally includes atoms and anything made up of them, and any particles (or combination of particles) that act as if they have both rest mass and volume.

(Excerpt end)

1.2 Standard Model

There has been a standard model for an atom for a long time. The Large Hadron Collider, LHC, has been used frequently to learn about the subatomic particles declared to be part of the Standard Model.

Excerpt from Wikipedia:

The current state of the classification of all elementary particles is explained by the Standard Model, which gained widespread acceptance in the mid-1970s
after experimental confirmation of the existence of quarks. It describes the strong, weak,
and electromagnetic fundamental interactions, using mediating gauge bosons. The species of gauge bosons are eight gluons, W−, W+ and Z bosons, and
the photon. The Standard Model also contains
24 fundamental fermions (12 particles and their associated anti-particles), which are the constituents of all matter.

(Excerpt end)

Observation:

The Standard Model has "widespread acceptance."

The standard model when observed from the context of cosmology becomes trivia, of interest only to only those in the narrow field of particle physics.

It is impossible for anyone lacking a LHC to duplicate its experiments or to confirm any conclusions drawn from using the LHC.

Expensive projects like LHC serve a small community of certain scientists.

Quarks are not relevant to this book's topic of gravity.

The standard model offers no viable mechanism for mass defect in atoms.

1.3 Structured Atomic Model

Recently, a new model for the atom has been proposed, the structured Atomic Model, or SAM. This model is important in understanding the foundation of matter.

A video presentation is found by:
"Edwin Kaal: The Proton-Electron Atom — A Proposal for a Structured Atomic Model | EU2017"

Excerpt from its header:

Importantly, this model does not contradict the evidence in chemistry and physics, but rather provides a new interpretation and a promisingly fresh approach. With this model, Edwin has been able to resolve enigmas in chemistry and make predictions to inform future research.

(Excerpt end)

Excerpt from the Structured Atomic Model web site:

Scientists believe there are protons and neutrons inside the nucleus of the atom. However there is much disagreement as to how it may be organized or even if it has an organization. Quantum mechanics states we can only understand the nucleus through mathematics and that the location of nucleons can have a range of locations, we will never know of a structure. Is this a reasonable conclusion? We know there are particles in there, it stands to reason they are arranged in some manner. Is it possible we just haven't discovered it yet?

The Structured Atom Model is a theory that the nucleus is highly organized and stable. This organization is responsible for the properties of the elements and chemistry. The rules which define how the nucleus is built are relatively simple and adhere to key principles found in nature.

(Excerpt end)

Observation:
To put it simply, the nucleus in the structured atomic model consists of only protons and electrons. A neutron is a temporary bond between the 2. When this pair leaves the nucleus, the bond dissolves leaving the pair in a few minutes. This is a known behavior for a neutron.

The difference in this model is there can be other particle combinations within a nucleus, like the alpha particle. This structure enables an explanation for the alpha particle emission during radioactive decay. This decay step is awkward for a nucleus having a random arrangement of its subatomic particles, as implied in the Standard Model.

SAM is an important recent development in chemistry and particle physics.

This book does not have to copy content about SAM. A link to the reference is in References at the end of the book.

SAM is important because SAM believed atomic matter can be explained using only 2 fundamental particles, electron and proton.

The structure of the protons and neutrons drives the properties of the atom, including the configuration of its electron orbitals.

However these orbitals really exist, absorption and emission lines are observed whenever electrons change their energy state in the atom.

However, from the author's research into SAM it offers no mechanism for mass defect in atoms.

The author started with a simple particle model as the basis for how only protons and electrons can explain the observed measured mass with atomic matter, including a mass defect. The simple beginning can be improved where necessary.

1.4 Definition of the atomic mass unit

The current atomic mass unit definition has a recommended change in this book.

This change can affect the claim of a mass defect, where the sum of the particles in an element does not add up to the element's measured atomic mass.

Some definitions from Wikipedia:

The dalton or unified atomic mass unit (symbols: Da or u) is a unit of mass widely used in physics and chemistry. It is defined as 1/12 of the mass of an unbound neutral atom of carbon-12 in its nuclear and electronic ground state and at rest. The atomic mass constant, denoted mu, is defined identically, giving $m_u = m(^{12}C)/12 = 1$ Da.

By definition, the mass of an atom of carbon-12 is 12 daltons, which corresponds with the number of nucleons that it has (6 protons and 6 neutrons). However, the mass of an atomic-scale object is affected by the binding energy of the nucleons in its atomic nuclei, as well as the mass and binding energy of its electrons.
Therefore, this equality holds only for the carbon-12 atom in the stated conditions, and will vary for other substances. For example, the mass of one unbound atom of the common hydrogen isotope (hydrogen-1, protium) is 1.007825032241 Da, the mass of one free neutron is 1.00866491595 Da, and the mass of one hydrogen-2 (deuterium) atom is 2.014101778114 Da.

In general, the difference (mass defect) is less than 0.1%; exceptions include hydrogen-1 (about 0.8%), helium-3 (0.5%), lithium (0.25%) and beryllium (0.15%).

1 u or 1 Da = $1.66053906660 \times 10^{-27}$ kg
1 1 u = $1822.888486209 \, m_e$

1 u = $1822.888486 \, m_e$
m_p = proton mass = 1.007276466621 u

e = electric charge = $1.602176634 \times^{-19}$ C
proton charge = +1e
m_e = mass electron = $5.48579909070 \times 10^{-4}$ u
electron charge = -1e

neutrino mass = $< 2.14 \times 10^{-37}$ kg, 95% confidence level, sum of 3 flavors

neutrino charge = 0e

neutrino mass = $< 3.53 \times 10^{-10}$ u

(Excerpt end)

Observation:

Data came from several Wikipedia topics.

The neutrino mass is my calculation using the kg to dalton conversion. It should be in the list here because it is a known subatomic particle though poorly understood, having an uncertain mass.

Excerpt from Wikipedia:

In physics, the proton-to-electron mass ratio, μ or β, is simply the rest mass of the proton (a baryon found in atoms) divided by that of the electron (a lepton found in atoms). Because this is a ratio of like-dimensioned physical quantities, it is a dimensionless quantity, a function of the dimensionless physical constants, and has numerical value independent of the system of units, namely:
$\mu = m_p/m_e = 1836.15267343$.

(Excerpt end)

Observation:

$1/\mu = 5.4462 \times 10^{-4}$

m_e uses this value and the proton mass.

At this point, the integrity of these assigned values could be checked.

However, the definition of 1 dalton does not provide the ^{12}C mass which was used for the calculation.

The particles in the ^{12}C atom, of 6 protons, 6 electrons, and 6 neutrons, which are each a proton and electron pair (by the Structured Atomic Model), can be summed with the result of 12.0873176 u

This is from:
 12 times 1.007276466621 u for 6 protons and 6 neutrons
+
12 times $5.48579909070 \times 10^{-4}$ u for sum of 6 neutrons
plus 6 pairs of a proton and orbiting electron.

It is impossible to check this result with certainty because Wikipedia did not state explicitly the value for ^{12}C being used in its Dalton definition.
Its current measured value in the Carbon topic is 12.0096 so this is assumed.
If the 12.0096 value was actually used for calculating 1 Dalton then that use was a mistake.

In the later section for Mass Defect, this value of 12.0096 appears to come from the isotope mix for carbon, which is not the correct benchmark for this critical calculation. The value used must be published to verify a correct calculation.
The mass of protium, or ^{1}H, is provided and will be used.

^{1}H = 1.007825032241

This atom is simply 2 particles:

$^{1}H = m_p + m_e$.

Using the two individual values the result is
1.007825046530

My Excel value is slightly higher than from Wikipedia.

The current masses of am electron and proton do not add up to the mass in a protium atom.

There can be no other reason for this difference than the values are wrong.

The protium (^{1}H) mass calculation can be from a different calculation using only one particle mass not two:
$$^1H = \mu\, m_e + m_e$$

or

$$^1H = (\mu + 1)\, m_e$$

This equation requires a high level of certainty of the precision of both the μ value and the ^{1}H value.

.
This result is 1.007825046538

This is not the measured value so either ^{1}H is wrong or m_e is wrong.

The m_e can be calculated with:

$$m_e = {}^1H / (\mu + 1)$$

with $^1H = 1.007825032240$ (Excel fails with last digit as 1

This is spec: $m_e = 5.48579909070 \times 10^{-4}$ u

The new result is $m_e = 5.485799012873 \times 10^{-4}$

Calculation using 10 digits, $m_e = 0.0005485799$

Though the last digit is dropped for Excel, the result was slightly higher, but this is a debatable number of significant digits for a valid comparison.

This is not the current m_e so either ^1H is wrong or μ is wrong – or the current m_e is wrong. This topic proposes m_e must change.

The m_p can be calculated using m_e and the ^1H spec value:

$$m_p = {}^1H - m_e$$

With the calculated m_e value, m_p = 1.007276452331
Or 1.0072764523 with only 10 decimal digits

Compare with result from ^{12}C: 1.007276466621

This is the result with the new m_p and m_e ^1H= 1.007825032232
Compare with this spec value: ^1H = 1.007825032241

The 2 new values sum to a slightly lower ^1H by only

Both the old pair and the new pair add up to slightly less than the current atomic mass value, but beyond the significant digits.

My Excel 2003 handles up to 10 digits after the decimal point. The numbers add up with that precision.

I cannot define a new mass for an electron with a suitable number of significant digits, if more than 10 are required.

The atomic mass values for the elements are rarely, if ever specified with more than 10 digits after the decimal.

This book's goal is to be practical.
The current mass values are probably wrong by a tiny amount. I expect there must be an agreement involving many contributors for such a change in the 2 fundamental particles.

This book uses a value but if this recommendation is accepted, then a value with a defined precision, or number of significant digits, must be agreed upon by those managing the "official" values.

This is the conclusion when trying to get the correct amu value:

This simple exercise using only ^{1}H and μ indicates physicists must confirm both values to the required precision, before assigning a mass value to the electron and proton if this alternate baseline is used instead of ^{12}C.

The author recommends this confirmation is a requirement before anything can be done with addressing the fundamental assumptions driving the atomic mass values of the two fundamental particles, the electron and proton.

The precision of the two crucial input values affects the precision of the resulting particle mass values.

With the protium as the reference not ^{12}C and getting different results indicates either:

a) assumed ^{12}C mass is wrong, or
b) protium mass is wrong, or
c) the value of μ is wrong, or

d) the specified values based on ^{12}C are wrong, or
e) there is a problem with the large number of significant digits for this analysis.

It is simply impossible for there to be "nuclear binding energy" in a nucleus consisting of only a proton.

Using the protium atom should be a better choice for defining the atomic mass unit because:

a) it consists of only the 2 fundamental particles,
b) it does not have18 particles (6 x proton, 6 x electron, 6 x neutron) like ^{12}C,
b) it has no possible binding energy,
c) it is not clear whether binding energy is accommodated in the ^{12}C algorithm,
d) it makes sense to use the unbreakable electron as the benchmark for defining atomic mass,

e) it is consistent with the structured atomic model treating the electron and proton as the fundamental particles,

f) the issue with this selection is it requires an accurate proton-to-electron mass ratio,

g) the precision of m_e depends on the precision of only ^{1}H and μ
The recommendation is a change to ^{1}H should be considered "again."

The amu has a history worth noting, described in this story:

Atomic Mass Unit Definition (AMU)

Excerpt:

John Dalton first suggested a means of expressing relative atomic mass in 1803. He proposed the use of hydrogen-1 (protium). Wilhelm Ostwald suggested that relative atomic mass would be better if expressed in terms of 1/16th the mass of oxygen. When the existence of isotopes was discovered in 1912 and isotopic oxygen in 1929, the definition based on oxygen became confusing.
Some scientists used an AMU based on the natural abundance of oxygen, while others used an AMU based on the oxygen-16 isotope. So, in 1961 the decision was made to use carbon-12 as the basis for the unit (to avoid any confusion with an oxygen-defined unit). The new unit was given the symbol u to replace amu, plus some scientists called the new unit a Dalton. However, u and Da were not universally adopted. Many scientists kept using the amu, just recognizing it was now based on carbon rather than oxygen.
At present, values expressed in u, AMU, amu, and Da all describe the exact same measure.

(Excerpt end)

Observation:

Stating just "Wilhelm Ostwald suggested" does not provide his reason for it being "better." The subsequent discovery of isotopes indicated it probably was not better.

After trying protium first, then oxygen, then carbon, one can suggest protium should have remained the standard. It would be awkward to change the benchmark element for a third time, to return to the original choice.

A hydrogen-2 (deuterium) atom can be checked with the new particle masses because ^{2}H mass is provided. Its nucleus is a proton and neutron.

Measured: ^{2}H = 2.014101778114

^{2}H = m_p + m_e + (m_p + m_e)
Or it is twice ^{1}H
To expect: t 2.0156500644

This is more than the specified mass.

When using the old calculated values:

Measured ^{2}H = 2.015650093

The differences are tiny, but notable. There is the expected binding energy between the nucleons.

With new particle masses, the difference is -0.000000029

Conclusion:

There is a known mass defect with the deuterium atom using the current particle masses. This is with either pair of values of electron and proton mass, based on either ^{1}H or ^{12}C.

This is the expected result because the measured is less than the sum, resulting in a mass deficit, which is called nuclear binding energy. The proton is binding with the neutron.

This comparison has 2 alternate explanations:
1) a neutron exhibited a loss in mass.
2) a neutron exhibited a loss in its reactivity to other masses.

This book suggests the second. No mass is becoming energy.

The difference between sets for ^{1}H ^{12}C can be compared for their summation for ^{12}C.

^{12}C is measured at 12.0096

Using the respective values for m_p and m_e, the results are:

From spec values ^{12}C = 12.09390056

From ^{1}H values ^{12}C = 12.0939004

There is a very small difference in the calculated values, beyond the number of significant digits (4 after decimal point) in the ^{12}C value.

However, the values which were supposed to result in exactly 12.0096 but their sum clearly failed to do so.

Both sets exhibit a mass defect because they do not match the measured atomic mass value,
The proton and electron mass values, derived from ^{12}C failed to result in exactly 12.0096. Because that was the goal of that algorithm, the algorithm failed.

That observation leads to a recommendation to use protium, which is the only atom having no possible binding energy.

Conclusion of atomic mass analysis:

This exercise suggests the current proton mass value is a tiny bit high.

The value derived from protium might be closer to the "correct" value if this is the only way to calculate the mass of a proton and electron.

One could suggest using protium, the simplest atom as the basis rather than carbon-12. The mass differences are very small.

Instead of recommending a change, the current values for the mass of an electron can remain unchanged, for now.

The big change in physics is explaining the observed mass defect.

Mass defect is currently explained as an awkward mass to energy conversion associated with Einstein.

One of the goals of this book is removing relativity from physics, so after mass defect is explained without relativity than another link is removed.

By the slight reduction in the proton mass, the mass defect in the protium atom is removed, as it should be.

As a result ALL elements will have a small reduction in their calculated mass defect.

Mass defect is detailed in the next section.

1.6 My simple subatomic particle model

This section describes the basis for the subsequent section about a new mechanism for gravity.

The few fundamental subatomic particles have a simple definition: mass and charge.

This model sometimes references the Structured Atomic Model (SAM) having only electrons and protons, is not based on the Standard Model having its zoo of quarks and quasi-particles.
The current definition of atoms including the configuration of electron shells is unchanged. The only tiny change in the atom is in the mass of electron and proton.

When ignoring the Standard Model, which treats quarks as fundamental particles:

There are only 6 fundamental subatomic particles, but from only 3 pairs when including rare anti-particles.

1) electron
   mass = m_e, charge = -1e

2) anti-electron or positron
   mass = m_e, charge = +1e

3) proton
   mass = m_p, charge = +1e

4) antiproton
   mass = m_p, charge = -1e

5) neutrino
mass = mass neutrino, charge = 0e

6) antineutrino
mass = mass neutrino, charge = 0e

The difference between neutrino and antineutrino is described by an attribute called chirality.

Chirality is not relevant to the topic of gravity. The antineutrino is in the list only because it is part of standard radioactive decay descriptions (below).

In this list, every particle has mass. Of the 3 pairs, there is only one particle pair, a neutrino, having no charge.

In a later section, I am proposing the force of gravity is distinct from an electric force which is driven by an electric charge.

A neutrino exhibits a "mass" behavior but no "charge" behavior. This particle confirms that assumption.

Every particle has mass while charge is sometimes present.

This distinction between mass and charge is important for particle pair production.

Maxwell defined behaviors of electric and magnetic fields, regardless of a particle's mass. Light involves these two fields but with no mass.

Newton defined a simple behavior between masses, regardless of their charge.

The amount of mass in a particle determines how strongly it reacts to the presence of other masses.

Mass reactivity involves generating a field which other masses react to, just like charges react to other charges.

The strength of this effect of mass is reduced by free space, so the effect is weaker with increasing distance. In the terminology of an electric field, the density of the field lines diminishes with increasing distance from the source.

The amount of charge in a particle determines how strongly it reacts to the presence of other charges.

A charge has different behaviors than a mass.

A moving charge creates what is called a magnetic field, with its strength determined by the amount of charge moving together.

A charge does not react to that magnetic field.

If the magnetic field is changing then it creates an electric field, so a charge will react to this change in its environment.

Observation about the fundamental particles:

A neutrino has very little mass so it reacts weakly to other masses.

The neutrino appears to never react to charges in its vicinity so it is assigned a value of zero charge.

This suggests internally the particle must have enough reactivity to other masses before it can react to other charges.

When a particle has the capability to react to other charges, every particle can have only one measured value of that reactivity, what is called the charge of an electron.

Having only one value implies there is a single internal mechanism driving its reactivity to other charges. This mechanism is active or not in each particle, and is determined at the moment of the particles creation. When the particle has this reactivity to charges, it has only one level of reactivity but 2 possible states.

This state, or polarity, either "+" or" –", is selected at the moment of particle creation and seems to be permanent.

This apparent permanence of polarity for an electron will be questioned below during particle pair production.

Its charge polarity is either positive or negative. When positive, it will be repelled by a net positive in its environment or attracted to a net negative in its environment. When negative, the opposite reaction occurs. This reaction is based on the polarity of the electric field lines in its environment.

There is no observed behavior for the "charge of an electron" to change from that specific value.

1.7 Comparison of forces

Atoms have charged particles.
This is a comparison of those forces.
The neutron has 2 distinct particles of opposite charge,
They are bound by both an electric force and a gravity force.

Values are from Wikipedia.

These equations are described again later.

Force of gravity equation:

$$F_g = G * (m1 * m2) / r^2$$

$$G = 3.67430 \times 10^{-11} \, m^3 \cdot kg^{-1} \cdot s^{-2}$$

$$e = \text{electric charge} = 1.602176634 \times 10^{-19} \, C$$

$$m_e = 9.1 \times 10^{-31} \, kg$$

$$m_p = 1.67262192369 \times 10^{-27} \, kg$$

$$radius_e = 2.8179403227 \times 10^{-15} \, m$$

$$radius_p = 0.84 \times 10^{-15} \, m$$

The electric force equation:

$$F_e = ke * (q1 * q2) / r^2$$

$$ke = 8.99 * 10^9 \, N \cdot m^2 \cdot C^{-2}$$

For the ^{1}H atom:

q1 or q2 = e
m1 = m_e
m2 = m_p

^{1}H atom radius = Bohr radius = 5.29 x $\times 10^{-11}$ m

So for proton and electron, in ^{1}H atom:

The gravity force:

F_g = 1.9795009170 x $\times 10^{-47}$ N

The electric force:

F_e = -8.2464899708$\times 10^{-8}$ N

F_e is negative meaning the force is attractive between opposite polarities.

When comparing F_e to F_g, the electric force is roughly 2.27×10^{39} stronger than the force of gravity.

The neutron will be mentioned later, but here is the appropriate place for this calculation and observation.

The neutron is usually in a nucleus. It disintegrates in a few minutes, when outside a nucleus.

So for proton and electron, in a neutron with radius$_p$ as their distance:

The gravity force:

$F_g = 7.2468016237E\text{-}23 \text{ x } \times 10^{-23}$ N

The electric force:

$F_e = -1.8866878216 \text{ x } \times 10^{-17}$ N

Protons will be adjacent to other protons or nearly so, in an atom's nucleus.

Using double the proton radius as the distance between their centers, the electric force and the force of gravity can be calculated.

The gravity force:

$F_g = 1.9795009170 \text{ x } \times 10^{-47}$ N

The electric force:

$F_e = +81.7639597476$ N
F_e is positive meaning the force is repulsive between the same polarity (positive) particles.

The amount of force in Fe is substantial.

This large value should be noted, because an alpha particle ejection will be described later.

The strong force within a nucleus must overcome this repulsion among the protons in near proximity, or even adjacent.
This repulsive force must be overcome during the process of fusion, where nucleons are added to a nucleus.

The electric and gravitation forces were described above, but in an atomic nucleus, the strong force dominates over the other forces.

Continue with other forces, including the strong force. Excerpt from Wikipedia:

In nuclear physics and particle physics, the strong interaction is the mechanism responsible for the strong nuclear force, and is one of the four known fundamental interactions, with the others being electromagnetism, the weak interaction, and gravitation. At the range of 10^{-15} m (1 femtometer), the strong force is approximately 137 times as strong as electromagnetism, a million times as strong as the weak interaction, and 1038 times as strong as gravitation.

In the context of atomic nuclei, the same strong interaction force (that binds quarks within a nucleon) also binds protons and neutrons together to form a nucleus. In this capacity it is called the nuclear force (or residual strong force). So the residuum from the strong interaction within protons and neutrons also binds nuclei together. As such, the residual strong interaction obeys a distance-dependent behavior between nucleons that is quite different from that when it is acting to bind quarks within nucleons.

Additionally, distinctions exist in the binding energies of the nuclear force of nuclear fusion vs nuclear fission.

(Excerpt end)

Observation:

Atomic nuclei "binding energies" will be described in the section titled Mass Defect.

Mass defect is a phenomenon which arises in the atomic nucleus.
This book is about gravity and how the mass in matter is measured.

A thorough explanation of the mechanism driving the strong force is out of scope for this book.

Condensed matter is another state of matter beyond the common states, like gas, liquid, solid and other rare states (like Bose-Einstein condensate).

Condensed matter has an atomic nucleus at every node in a lattice, which is maintained by free electrons. When protons are at the nodes of the lattice, this atomic structure involving protons is called metallic hydrogen.

Excerpt from Wikipedia:

Condensed matter physics is the field of physics that deals with the macroscopic and microscopic physical properties of matter. In particular it is concerned with the "condensed" phases that appear whenever the number of constituents in a system is extremely large and the interactions between the constituents are strong.

The most familiar examples of condensed phases
are solids and liquids, which arise from
the electromagnetic forces between atoms.

Condensed matter physicists seek to understand the
behavior of these phases by using physical laws. In
particular, they include the laws of quantum
mechanics, electromagnetism and statistical mechanics.
More exotic condensed phases include
the superconducting phase exhibited by certain materials
at low temperature,
the ferromagnetic and antiferromagnetic phases
of spins on crystal lattices of atoms, and the Bose–Einstein
condensate found in ultracold atomic systems.

The study of condensed matter physics involves
measuring various material properties via experimental
probes along with using methods of theoretical physics to
develop mathematical models that help in understanding
physical behavior.
The diversity of systems and phenomena available for
study makes condensed matter physics the most active
field of contemporary physics: one third of
all American physicists self-identify as condensed matter
physicists, and the Division of Condensed Matter Physics
is the largest division at the American Physical Society.

(Excerpt end)

There is more information on condensed matter below.

A force similar to the strong force must overcome this
mutual repulsion among the positive nuclei to maintain the
lattice.

The strong force and an explanation of condensed matter are not part of this book's topic, the definition of a force of gravity.

Our Sun is composed of much liquid metallic hydrogen. The planet Jupiter is also assumed to have much of its mass in the form of liquid metallic hydrogen.

The accurate compositions of the Earth and other celestial bodies have a level of uncertainty.

1.8 Neutron

The neutron will be mentioned again in the mass defect section, but here is the appropriate place for this observation.

Excerpt from Wikipedia for Neutron:

The neutron is a subatomic particle, symbol

n

or

n^0

, which has a neutral (not positive or negative) charge and a mass slightly greater than that of a proton. Protons and neutrons constitute the nuclei of atoms. Since protons and neutrons behave similarly within the nucleus, and each has a mass of approximately one atomic mass unit, they are both referred to as nucleons. Their properties and interactions are described by nuclear physics.

(Excerpt end)

Observation:

This book assumes an atom consists of protons and neutrons in the nucleus with electrons in orbit around it. The behavior of quarks in the standard model are never in this author's explanations.

A neutron is created by a tight bond between a proton and an electron.
The expected mass of a neutron is the sum of its 2 particles.

Neutron mass = $m_p + m_e$

Using current particle masses:

The result is 1.0078250465

The current measured mass of a neutron is 1.0086649159

The difference is 0.000839869350

Note the sum is greater than the measured.

The standard model accepts a neutron's measured mass is greater than the sum of its 2 particles.

This difference is called a mass defect.

The author proposes a slight change to m_p and m_e
So their sum exactly matches the mass of the protium atom which has only those 2 particles.

The result is 1.0078250322

The difference is 0.0008398836

Note the sum is greater than the measured, like with current values.

The sum of the two is greater than that of a neutron. This suggests the reactivity to other masses decreased while the reactivity to other charges neutralized by the pair because the neutron is assigned a charge of zero.

However, the neutron's mass was measured indirectly using a deuterium atom so its accuracy could be questioned before making this conclusion.

An electron has a fixed mass and a fixed charge.
An electron never forms a bond with another electron.
Protons can bind with another proton in a nucleus. Every bond involves a proton.

Various combinations in a nucleus will be described later.

It is reasonable to conclude the decrease in reactivity to another mass, during a bond between particles, occurs in the proton, not the electron. The measured mass defect is usually more than the mass of an electron, usually a multiple.
The disintegration of a neutron results in a set of a proton and electron pair and a neutrino having an unmeasured mass.

This result indicates the 2 charge reactivity components of opposite polarity remain intact while observed as a neutron.

This explains its net charge of zero, and the separate polarities are maintained before and after disintegration.

From the descriptions, there is no release of energy at the moment of disintegration, but only the three particles.

The origin of the neutrino is not always clear, though several scenarios are offered.

Mass seemed to be lost initially but restored later, at the same moment a tiny mass neutrino appears.

Therefore, the initial two particles apparently persist as individual particles.

The Standard model describes a neutron as 3 separate quarks, not the pair of proton and electron.

Image and caption from Wikipedia:

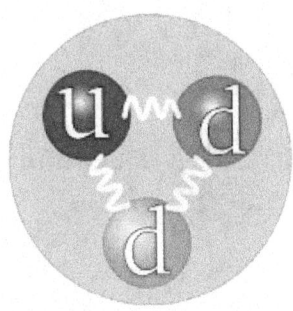

The quark content of the neutron. The color assignment of individual quarks is arbitrary, but all three colors must be present. Forces between quarks are mediated by gluons.

1 up quark, 2 down quarks

(Caption end)

Observation:

This representation is appropriate when comparing atomic models.

The disintegration of a neutron is a known behavior.

Neutrinos are uncharged particles having a tiny mass. Neutrinos were mentioned in section 1 of this book.

A neutrino is emitted when a neutron disintegrates into a proton-electron pair. This process is also called beta decay because the electron is also called a beta particle.

The Sun is known to emit neutrinos, so they are assumed to originate from the fusion or transmutation activity there.

Therefore the image implying a neutron consists of only 3 fragments and not 2 distinct particles disagrees with observations.

Excerpt from Wikipedia:

Research is intense in the hunt to elucidate the essential nature of neutrinos, with aspirations of finding:

- the three neutrino mass values
- the degree of CP violation in the leptonic sector (which may lead to leptogenesis)
- evidence of physics which might break the Standard Model of particle physics, such as neutrinoless double beta decay, which would be evidence for violation of lepton number conservation.

(Excerpt end)

Observation:

Neutrinos are not well understood and are not so important in this book about gravity. They have a tiny reactivity to other masses.

1.8.1 Neutron's mass

One might expect the mass of a neutron is measured directly. That is not the case. It uses the assumption of nuclear binding energy which this book suggests is not really as claimed.

Excerpt from Wikipedia:

The mass of a neutron cannot be directly determined by mass spectrometry due to lack of electric charge. However, since the masses of a proton and of a deuteron can be measured with a mass spectrometer, the mass of a neutron can be deduced by subtracting proton mass from deuteron mass, with the difference being the mass of the neutron plus the binding energy of deuterium (expressed as a positive emitted energy). The latter can be directly measured by measuring the of the single 0.7822 MeV gamma photon emitted when neutrons are captured by protons (this is exothermic and happens with zero-energy neutrons), plus the small recoil kinetic energy) of the deuteron (about 0.06% of the total energy). The energy of the gamma ray can be measured to high precision by X-ray diffraction techniques, as was first done by Bell and Elliot in 1948. The best modern (1986) values for neutron mass by this technique are provided by Greene, et al. These give a neutron mass of:
[mass] neutron = 1.008644904 u
The value for the neutron mass in MeV is less accurately known, due to less accuracy in the known conversion of u to MeV:
[mass] neutron = 939.56563 MeV/c^2.

Another method to determine the mass of a neutron starts from the beta decay of the neutron, when the momenta of the resulting proton and electron are measured.

(Excerpt end)

Observation:

There are many assumptions to get the precise result, including a recognized issue with the "less accuracy" of the conversion from a gamma ray wave length to its equivalent mass.

The mass defect behavior appears driven by neutrons. The measured mass of a neutron was calculated by using the atom with a proton and neutron in its nucleus, so the conditions inherently involved a mass defect, which is a behavior lacking a good explanation in the standard model.

This technique is suspicious.

This brings into doubt whether the measured mass, defined with much precision, is actually verified to that accuracy.

Calculations using many significant digits must beware factors lacking the precision of other values when claiming the final precision.

The neutron behaves like 2 distinct particles but when in a nucleus, it can exhibit a reduced reactivity to other masses. This is the conclusion after the comparison between the 2 atoms having 3 nucleons but different sets.

The 2 distinct charge behaviors appear intact but the reactivity to other masses is changed slightly only while bonded.

When the split occurs, each charge components gets its original mass reactivity component.
The split is accompanied by a neutrino having little mass reactivity and no charge reactivity.

When an electron and proton unite to form a neutron, the mass reactivity, from the accepted value) is less than the sum.

This confirms the mass behavior is not just an addition. The mass reactivity of a neutron is not driven by the sum at its creation. In other words, a proton can exhibit a difference in its expected mass while in a nucleus. This is called a mass defect. Mass defect has its own section in this book.

1.9 Quarks

Particle accelerators have been unable to break an electron so it remains a fundamental particle regardless of a particular atomic model.

When smashing either a proton or neutron into pieces, the same result is 3 unknown particles which are not found outside a collider.

Apparently, the 3 remnants exhibit no charge but they must exhibit mass to be detected,

The sets of three have different names because the proton had charge while the neutron had none, so the assumption is the sets must be different in some way.

One possible scenario is the mass reactivity split into 2 while the original charge behavior was the third particle with a weak mass reactivity and its charge behavior disabled or "broken" upon impact.

This book is about atomic matter, and the 2 fundamental particles, not their fragments.

1.8 Antimatter.

An antiproton is never found in the universe. One can be created only in a particle accelerator.

From Wikipedia:
"The antiproton was first experimentally confirmed in 1955 at the Bevatron particle accelerator by University of California."

The antiproton can be ignored in this book.

A positron can be found after 2 transient events.

1) positron emission,

2) particle pair production.

Excerpt from Wikipedia for (1):

Positron emission or beta plus decay (β+ decay) is a subtype of radioactive decay called beta decay, in which a proton inside a radionuclide nucleus is converted into a neutron while releasing a positron and an electron neutrino (ve). Positron emission is mediated by the weak force. The positron is a type of beta particle (β+), the other beta particle being the electron (β−) emitted from the β− decay of a nucleus.

Excerpt for (2):

Pair production often refers specifically to a photon creating an electron–positron pair near a nucleus. For pair production to occur, the incoming energy of the photon must be above a threshold of at least the total rest mass energy of the two particles, and the situation must conserve both energy and momentum. However, all other conserved quantum numbers (angular momentum, electric charge, lepton number) of the produced particles must sum to zero – thus the created particles shall have opposite values of each other. For instance, if one particle has electric charge of +1 the other must have electric charge of −1, or if one particle has strangeness of +1 then another one must have strangeness of −1.

The probability of pair production in photon–matter interactions increases with photon energy and also increases approximately as the square of atomic number of the nearby atom.

(Excerpts end)

Observation:

SAM describes a neutron as the triple combination of a proton, electron, and an electron antineutrino, where a neutrino is called the "activation energy."

Item (1) involves radioactive decay, which is not relevant to gravity.

Item (2) is relevant here.

Particle pair production probability increases with more electrons.

Every atom has a specific configuration of its electron orbitals.

An atom will absorb a specific wavelength when it can change its set of electrons to a new energy state by that amount being absorbed. This is a quantized behavior of an atom, where a longer wavelength, having less energy than required, will not be absorbed by the atom.

Similarly, when an electron moves to a lower orbital, or to a lower energy state, an emission line of a particular wave length is emitted. This wave length is sometimes related to the distance between orbitals and it contains the energy being released from the electron's change in its energy.

The photoelectric effect has an extra result with the absorption line, by ejecting an electron.

When the atom absorbs enough energy for an electron to leave the atom rather than just changing orbitals,
then the electron leaves having the kinetic energy with the excess over the minimum required to leave.

The pair production event description is awkward in the excerpt, with "[creating a] pair near a nucleus" when the event is actually changing an electron pair in orbit around the nucleus. Of course, that is "near."

This is an internet reference:

Planck's Constant and the Nature of Light

The YouTube video of that title by Lori Gardi is recommended for 3 reasons:

1) The well accepted Planck equation has a bug. She has a thorough explanation which is worthwhile to hear.

2) The energy in light is in the intensity of a particular wave length, not in the frequency, as is currently assumed.

3) This is another useful explanation of why there is no photon.

This book like a number of scientists do not agree the Standard Model is correct. One its quasi-particles is the photon. Light is a wave, not a particle.

This had been noted in the author's first book during its spectrum analysis.

The conclusion that wave length intensity carries the energy is relevant to some positron events.

Particle pair production event requires an atom, having multiple electrons, absorbing the energy in a very short wavelength, or in the gamma ray range.

The energy for the photoelectric effect results in an electron emission. Each atom has a minimum wave length defining the energy required for this effect. The wave length affects this ejection, not just intensity. These events are usually in the ultraviolet range.

The higher energy required for the pair production results in the emission of both an electron and a positron. This suggests the atom is losing 2 electrons but one flipped its charge polarity during its ejection.

The description mentions the energy matching the sum of the pair of particle masses, as if both were created just from energy.

1.9 A different mechanism for particle pair production

A simpler, alternative explanation of this event is the extra energy being absorbed is applied to flipping the charge polarity in one electron of the pair. Then there is no creation of mass in a new particle.

As noted earlier, each fundamental subatomic particle has mass. All but the neutrino has a charge.

Charge is a separate behavior in a particle from mass. However, each behavior is driven by energy, because each behavior drives a pervasive field, one for gravity and the other for electric, and each particle reacts to the pervasive fields from other particles.
Because the reactions are instantaneous, the results are also instantaneous.

It is simpler to propose the mechanism for a charge to flip polarity than it is to propose two new particles with consistent attributes get created from energy.

It is not clear whether the absorbed energy always exactly matches the particle pair. The energy is distributed between the ejection and kinetic velocity of the ejected particles, and perhaps that required for the polarity change.

There is no known mechanism to create an electron or proton. It is inconsistent proposing there is a special mechanism only for one antiparticle, the positron. The electron in the particle pair is coming from the atom.

The photoelectric effect ejects one particle while pair production ejects two. Neither event actually creates matter.

A muon subatomic particle is observed with some particle accelerator experiments. Its mass is about 207 times that of an electron and also has a charge of -1e. Its exact mass is

0.1134289259 Da.

When compared to other fundamental particles, different masses never change the amount of charge, an intrinsic property.

Results of particle accelerators, like a muon, quarks, other subatomic fragments, or quasi-particles, are not relevant to gravity which is the topic of this book.

There is this combination:

a) Muon has a mass of "about" some multiplier of an electron.

b) Neutrinos apparently have different masses among its types (electron, muon, tau).

One could conclude there is no defined intrinsic increment for a particle's reactivity to other masses, though the ratio between masses of electron to proton is known to many significant digits.

An observation on antimatter annihilation is possible.

The collision of a particle with its anti-particle, like electron with positron, releases much energy.
When colliding, there are 2 charged particles in motion for a brief moment, over a very short distance during a mutual annihilation. Charges in motion during a change an atom's energy state can result in radiation as emission lines, defining its characteristic wave length.

When the distance is smaller than an electron and it defines the wave length emitted from this event, then the energy being released is carried in a tiny wave length or in gamma rays. One could suggest 2 wave lengths are emitted, one from each charged particle, for the energy within that particle being annihilated.

I have wondered how to explain the gamma ray wave length from extreme events, like anti-particle annihilation. Perhaps, that is one way. Sometimes during the radioactive decay of a complex nucleus, a gamma ray emission occurs. The distance during each event is tiny.

Excerpt of the conventional explanation of Electron–positron annihilation, from Wikipedia:

When a low-energy electron annihilates a low-energy positron (antielectron), the most probable is the creation of two or more photons, since the only other final state Standard Model particles that electrons and positrons carry enough mass-energy to produce are neutrinos which are approximately 10,000 times less likely to produce, and the creation of only one photon is forbidden by momentum conservation, [because] a single photon would carry nonzero momentum in any frame, including the center-of-momentum frame where the total momentum vanishes. Both the annihilating electron and positron particles have a rest energy of about 0.511 million electron volts (MeV). If their kinetic energies are relatively negligible, this total rest energy appears as the photon energy of the photons produced. Each of the photons then has an energy of about 0.511 MeV. Momentum and energy are both conserved, with 1.022 MeV of photon energy (accounting for the rest energy of the particles) moving in opposite directions (accounting for the total zero momentum of the system).

If one or both charged particles carry a larger amount of kinetic energy, various other particles can be produced. Furthermore, the annihilation (or decay) of an electron-positron pair into a single photon can occur in the presence of a third charged particle to which the excess momentum can be transferred by a virtual photon from the electron or positron. The inverse process, pair production by a single real photon, is also possible in the electromagnetic field of a third particle.

(Excerpt end)

Observation:

It is not explicit but probably there are individual sort wave lengths emitted in opposing directions. This result was suggested in the text above the excerpt.

This excerpt is awkward after watching the video about Planck's constant. There are no photons and energy is carried in the intensity of the particular wave length.

The excerpt's use of "can be" reveals some of this is conjecture.
Virtual photons are used as an intermediate state within a quantum sequence but with no definition of that state.

Excerpt from Britannica:

QED rests on the idea that charged particles (e.g., electrons and positrons) interact by emitting and absorbing photons, the particles that transmit electromagnetic forces. These photons are "virtual"; that is, they cannot be seen or detected in any way because their existence violates the conservation of energy and momentum. The photon exchange is merely the "force" of the interaction, because interacting particles change their speed and direction of travel as they release or absorb the energy of a photon. Photons also can be emitted in a free state, in which case they may be observed as light or other forms of electromagnetic radiation.

The interaction of two charged particles occurs in a series of processes of increasing complexity. In the simplest, only one virtual photon is involved; in a second-order process, there are two; and so forth. The processes correspond to all the possible ways in which the particles can interact by the exchange of virtual photons and each of them can be represented graphically by means of the so-called Feynman diagrams. Besides furnishing an intuitive picture of the process being considered, this type of diagram prescribes precisely how to calculate the variable involved. Each subatomic process becomes computationally more difficult than the previous one, and there are an infinite number of processes.

The QED theory, however, states that the more complex the process—that is, the greater the number of virtual photons exchanged in the process—the smaller the probability of its occurrence.
(Excerpt end)

Observation:

The use of "can be" "probability" and "virtual" reveal some of this is just conjecture for "processes of increasing complexity."

The use of photons when such a particle does not exist is just compounding the confusion.

My conclusion:

Mass and charge are separate attributes of matter, down to the 6 fundamental subatomic particles, and the muon.

Many agree: a photon does not exist.

I suggest a virtual photon also does not exist.

1.10 Fast Radio Burst

This topic is not related to gravity. However, particle pair production was described above that justifies its inclusion.

the title of a news story:

Mysterious 'fast radio burst' detected closer to Earth than ever before

Excerpt from the study:

This is the first burst with a radio counterpart observed from a soft γ-ray repeater and it strongly supports models based on magnetars that have been proposed for extragalactic fast radio bursts.

(excerpt end)

observation:
There is a "gamma ray repeater" here, associated with the Fast Radio Burst, or FRB.

When the energy of gamma rays is absorbed by an atom, particle-pair production can occur. This pair is often an electron and positron.

The result of this event is two slowly moving charged particles. When their motion is altered by a magnetic field, they become a source of synchrotron radiation, but with a slow velocity, the peak wave length is down in the radio frequency range.
The radio burst continues as long as these particles persist.

The radio burst ends when the electron and positron pairs were mutually annihilated.

The story mentions a magnetar. That is a plasmoid, an entity described in the section about black holes. A black hole is actually a plasmoid, which is capable of synchrotron radiation spanning frequencies from X-ray to radio. When its electric current is even more intense, that high end can increase from X-ray to gamma ray.

1.11 Condensed Matter

Condensed matter is a lattice configuration of atomic nuclei held together by loose electrons. Graphite is an example of condensed matter where the carbon nucleus is each point in the lattice.

This lattice of many nuclei is not a form of a molecular bond. Molecules take a shape based on the nuclei in the bond. For example water is a compound of 1 oxygen atom and 2 hydrogen atoms which is not symmetrical giving water characteristics from that lack of symmetry.
A lattice must have symmetry in its structure.

Metallic hydrogen is another example of condensed matter but a proton is at every node in the lattice.

In several sciences, the word proton is avoided so hydrogen atom is used instead. Plasma physics arose with Hannes Alfven in 1970, but it has not been integrated into astrophysics. Alfven proposed a model for a spiral galaxy disk rotation. That was ignored, leading to the mistake of dark matter.

Metallic hydrogen has been proposed for the Sun. This theory became public with the Dr. Pierre-Marie Robitaille paper in 2013 titled:

"Forty Lines of Evidence for Condensed Matter - The Sun on Trial: Liquid Metallic Hydrogen as a Solar Building Block"

Dr. Robitaille has a YouTube channel, Sky Scholar, with many videos about stars and this model, along with other topics including thermodynamics.
Among the videos is:

The Sun is NOT a Gaseous Plasma! The LMH Solar Model!

This model matches the helio-seismology data and all the crucial solar observations, like limb darkening and the corona. The gaseous sun model lacks sufficient evidence, and it is unable to explain all observations, including its liquid surface, different rotation rates by latitude, its hot corona, solar wind acceleration, and many more.
An online reference for Condensed Matter Physics has a chart showing the lattice for each named structure:

Bravais lattices

Observation:

The metallic hydrogen lattice is only protons, maintained by free electrons.

Those protons should be mutually repulsive by their positive charge but the lattice is stable.
The lattice has a gap between particles so there should be no mass defect from the structure.

The lattice configuration changes with increasing depth in the Sun. The photosphere and convective zone are liquid metallic hydrogen. The solar core is solid metallic hydrogen with the body-centered cubic lattice.

Those changes in lattice density by depth suggest gravity is also a contributor to lattice stability.

The author's book Cosmology Transition included a description of the solar layers in this LMH model.

1.12 Solar Neutrinos

This section mentioned neutrinos and the solar model.

A comment on the combination is worthwhile.

Excerpt from Wikipedia about a neutrino:

[The] Homestake experiment made the first measurement of the flux of electron neutrinos arriving from the core of the Sun and found a value that was between one third and one half the number predicted by the Standard Solar Model. This discrepancy, which became known as the solar neutrino problem, remained unresolved for some thirty years, while possible problems with both the experiment and the solar model were investigated, but none could be found.

Eventually it was realized that both were actually correct, and that the discrepancy between them was due to neutrinos being more complex than was previously assumed. It was postulated that the three neutrinos had nonzero and slightly different masses, and could therefore oscillate into undetectable flavors on their flight to the Earth.

(Excerpt end)

Observation:

The Sun is not powered by fusion. The prediction of neutrinos failed because the solar model is wrong.

The scientific method steps include: if a theory's prediction fails then the theory must be fixed.

In this case, the theory remains unchanged, while the particles are claimed to change while in transit. That implies these neutrinos are unstable. There is no evidence by confirming neutrinos at their source are different than found here.

Actually, the element transmutation, or Low Energy Nuclear Reaction, LENR, occurs on the photosphere, not in the core.

This process has been confirmed by the SAFIRE project which created in a laboratory several controllable conditions to emulate the solar surface.

The liquid metallic hydrogen solar model, having no internal fusion, has been presented a number of times, including to the American Physics Society.

Such a fundamental change in astrophysics is difficult to achieve.

1.13 Particle Creation

There is no known mechanism to create matter.

Currently, the particle pair production explanation proposes a pair of electron and positron is created from one atom after it has absorbed enough energy from a wave length of electromagnetic radiation.

An alternate explanation, in section 1 of this book, is the electron in the pair is one from the atom while the positron is another of the atom's electrons which changed its polarity before being ejected.

With this explanation, neither electrons nor positrons are ever created by this process.
The universe has no mechanism to create mass in the form of electrons and protons.

The non-fusion model of the Sun was described earlier. This model is growing in acceptance. Fusion required the conversion of mass into energy, resulting in mass being lost from the universe. With an electrical mechanism for a star's energy, a star is not losing mass.
The black hole is described later in this book. Such a thing does not exist. It was claimed to remove matter from the universe and put it into a point in the special observer's reference frame, called a singularity. This does not occur so no matter is lost from the universe by this theoretical object.

Therefore, the universe is not actually losing mass as assumed by current cosmology with stellar internal fusion and black holes.

The author's first book, Observing Our Universe, revealed the big bang theory is just science fiction with no evidence to justify any aspect of the theory. There is no creation of matter after a catastrophic explosion, followed by a complicated sequence from an initial unknown singularity, to all the matter currently observed, with no verification for anything in that sequence.

The universe has its quantity of mass and is not adding to that quantity.

1.14 Radioactive Decay

This is not about gravity but the neutron has become a critical component in a nucleus.
Radioactive decay involves neutrons in several ways.

Excerpt from Wikipedia:

Radioactive decay (also known as nuclear decay, radioactivity, radioactive disintegration or nuclear disintegration) is the process by which an unstable atomic nucleus loses energy by radiation. A material containing unstable nuclei is considered radioactive. Three of the most common types of decay are alpha decay, beta decay, and gamma decay, all of which involve emitting one or more particles or photons.

Except for gamma decay or internal conversion from a nuclear excited state, the decay is a nuclear transmutation resulting in a daughter containing a different number of protons or neutrons (or both). When the number of protons changes, an atom of a different chemical element is created.

(Excerpt end)

Observation:

Several behaviors are involved in the different steps of decay.

One or more neutrons participate in any step.

The only radioactive element with a measured mass value exhibits a substantial mass defect.

That observation suggests radioactivity includes the nuclei having more than one anomaly at the same time.

Radioactivity, by itself is not a topic critical to this book. There is only one radioactive element having valid data for analysis.

However, radioactivity is mentioned again later in the new atomic model

1.15 Velocity Limits on Light and Mass

Einstein wrongly assumed mass had a velocity limit at c, the velocity of light in a vacuum.

There are various claims of a speed limit for light and for mass.

A YouTube video is titled:
Why is the speed of light what it is? Maxwell equations visualized

This video clearly explains the velocity limit for light but makes a drastic mistake when mentioning Einstein. Einstein is mentioned separately below

From Wikipedia topic on Maxwell's equations:

Maxwell's equations explain how these [light] waves can physically propagate through space. The changing magnetic field creates a changing electric field through Faraday's law. In turn, that electric field creates a changing magnetic field through Maxwell's addition to Ampère's law. This perpetual cycle allows these waves, now known as electromagnetic radiation, to move through space at velocity c.

Relative permittivity is the factor by which the electric field between the charges is decreased relative to vacuum.

(Excerpt end)

There are also factors for relative permeability and magnetic susceptibility.

Observation:

The propagation of light is a self-propagating series of electric and magnetic fields. Its velocity is determined ONLY by the medium. An instantaneous change in medium causes an instantaneous change in propagation velocity. This transition is observed at the surface of glass or water.

Conclusion:

This propagation begins and continues at the same velocity regardless of any velocity of the source of its propagation.

All of the above is correct, regardless of anything Einstein claimed.

He had nothing to do with the constant velocity of light.

Some question whether a moving light source affects the velocity of its light. Einstein had thought experiments about that, when he concluded only the special observer might see its velocity change.

The velocity of light propagation is always defined by the medium.

The velocity and direction of a light source affects the energy in the light but not its rate of propagation.

If the light source is moving then it has kinetic energy.

In thermodynamics, energy cannot be lost or gained but only exchanged or transformed.

Around the sphere of radiated energy, wave lengths are reduced in the direction of travel or increased in the opposite direction. Each wave length change is determined by the velocity and direction at that point relative to c, the constant velocity of light in a vacuum. If the light source is in a medium, the z is the same because the medium affects the propagation not the initiation. Energy is maintained around the radiated sphere at the instant of emission and the propagation velocity of the light emission is not affected.

The velocity and direction of a moving light source has no effect on the velocity of its light.

Einstein did have something to do with the velocity of mass, a mistake.

He worked only with the context of a moving observer, the "special" observer in both "general" and "special" relativity..

Einstein's belief, of a velocity limit on mass, was shared by others in the 1800's, but has no justification in physics.

For many years, Einstein's unjustified belief has been refuted.

Excerpt from Wikipedia:

In 1993, Thomson et al. suggested that the (outer) jet of the quasar 3C 273 is nearly collinear to our line-of-sight. Superluminal motion of up to ~9.6c has been observed along the (inner) jet of this quasar.

Superluminal motion of up to 6c has been observed in the inner parts of the jet of M87. To explain this in terms of the "narrow-angle" model, the jet must be no more than 19° from our line-of-sight.

(Excerpt end)

Observation:

A plasmoid, like in 3C 273 or M87, holds substantial electromagnetic energy. The sustained force of a magnetic field on charged particles can result in velocities far faster than c and this has been measured many times.

The motion of mass is not a process dependent on the medium.

Motion is affected by the medium only with its friction on the surface of the mass in motion. Friction is an exchange of kinetic energy to thermal energy. Friction does not define a velocity limit.

Applying a force to a mass results in its acceleration. The force can be maintained for a specific time to achieve the desired velocity. This is a continuous transfer of energy from the source to kinetic energy.

This process of energy transfer is observed during every launch of a space probe. Power is the amount of force during a time. The force required is determined by the mass and the time required for the desired velocity.

The available power is determined by the fuel supply, which is "full" at the moment of launch. The number and design of the respective stages determines the final velocity of the final stage which has the lowest mass, where individual stages provide the required amount of power for the velocity of the remaining stages.
Increasing the initial power can increase the final velocity.

Once a mass is in motion it has kinetic energy. It must maintain that energy, so it remains in motion, until this energy is transferred like with friction. Friction is a transfer of kinetic energy into thermal energy.

As long as a force continues to transfer more energy into more kinetic energy, the velocity must increase. There is no velocity limit during this energy transfer.

Einstein wrote equations causing changes to the moving observer when near the velocity of light, including relativistic mass, so their possible velocity limit was set in math not physics. At $c = v$, the relativistic mass is a mass divided by zero which is either infinite or not allowed!

From Wikipedia:
"Oxford lecturer John Roche states that relativistic mass is not referenced in nuclear and particle physics, and that about 60% of authors writing about special relativity do not introduce it."

Observation:

This unverified prediction by relativity is widely ignored, while some claim all predictions by relativity were confirmed.

That claim of relativistic mass is unverified and should be retracted for many reasons. Other problems with relativity are described in later sections in this book.

Proposing mass has a velocity limit at c means when applying more force to a mass when nearing or at the velocity of c, energy is being lost rather than transferred to kinetic energy.

This loss is a violation of thermodynamics.

The practical velocity limit for a specific mass is set by the power required to reach that velocity.

1.16 Conclusion on velocity limits

The velocity limit on light is set by the medium.

Light always propagates at a velocity set by the medium.

There is no defined velocity limit on mass.

2 Mass Defect

The phenomenon called mass defect requires a new explanation in physics.

It is not really a defect. It is just an observation of a different mass than expected.

At least the word "defect" is better than "dark."

Dark matter is proposed when there is apparently missing matter on a cosmological scale.

The cause of that missing matter is usually a magnetic field is missed or ignored. A magnetic field affects charged particles in motion. Nearly all matter in the universe is plasma, or it has a charge.

On the atomic scale, a difference in mass can be observed between expected and measured in an atomic nucleus. The difference can be more or less than expected. When it is less then expected, it can be called a mass deficit.
The nucleus consists of protons and neutrons, but each neutron is a combination of proton and electron. The electron is a tiny particle compared to a proton. The mass defect is an apparent change in the protons in the nucleus.

After this phenomenon has an explanation, then it should not be called a defect.

There are 2 sources to compare their descriptions.

Data for the elements in this section 2 are from Wikipedia, unless another source is identified. Links are in References.

2.1 Description

Mass defect has this description, from Britannica:

The observed atomic mass is slightly less than the sum of the masses of the protons, neutrons, and electrons that make up the atom. The difference, called the mass defect, is accounted for during the combination of these particles by conversion into binding energy, according to an equation in which the energy (E) released equals the product of the mass (m) consumed and the square of the velocity of light in vacuum (c); thus, $E = mc^2$.

(Excerpt end)

Observation:

A difference is not always explained by this mass/energy relationship.

Wikipedia uses another name and has no topic for Mass Defect.

In Wikipedia, the topic "mass defect" refers to an anomaly in a spiral galaxy brightness profile near its core.

Wikipedia calls the atomic mass defect behavior something else.

Excerpt from Wikipedia:

Nuclear binding energy is the minimum energy that would be required to disassemble the nucleus of an atom into its component parts. These component parts are neutrons and protons, which are collectively called nucleons. The binding energy is always a positive number, as we need to spend energy in moving these nucleons, attracted to each other by the strong nuclear force, away from each other. The mass of an atomic nucleus is less than the sum of the individual masses of the free constituent protons and neutrons, according to Einstein's equation $E=mc^2$. This 'missing mass' is known as the mass defect, and represents the energy that was released when the nucleus was formed.

Mass defect (also called "mass deficit") is the difference between the mass of an object and the sum of the masses of its constituent particles. Discovered by Albert Einstein in 1905, it can be explained using his formula $E = mc^2$, which describes the equivalence of energy and mass. The decrease in mass is equal to the energy given off in the reaction of an atom's creation divided by c^2. By this formula, adding energy also increases mass (both weight and inertia), whereas removing energy decreases mass. For example, a helium atom containing four nucleons has a mass about 0.8% less than the total mass of four hydrogen nuclei (which contain one nucleon each). The helium nucleus has four nucleons bound together, and the binding energy which holds them together is, in effect, the missing 0.8% of mass.

(Excerpt end)

The entire periodic table is reviewed in this section to find those elements with either "missing mass" or its opposite "extra mass."

The fact both occur among the elements clearly indicates the behavior is different than currently explained, as in only the case of "missing mass" replaced by energy.

In the opposite, the nucleus has somehow converted energy into mass.

This is a violation of thermodynamics when having an unidentified source of external energy.

2.1 Mass Defect Data Analysis

The author quickly noted the mass defect increases with a heavier atom. This suggests the number of nucleons is part of this behavior.

There are only a few elements exhibiting "extra mass" rather than "missing mass" so the neutrons in the nucleus are the likely components causing this change.

A spread sheet expedites this analysis.

Each element has a mix of entries and calculations.

There are only manual entries:

a) its defined number of protons,

b) its nominal atomic weight,

c) its measured atomic weight (with precision to 10 digits, if available).

Microsoft Excel handles up to this precision. That is sufficient for this book.

From these 3 entries, the number of neutrons is available, so the atom's particles can be summed. This is called the predicted mass.

The measured mass value subtracts the predicted mass to obtain the mass defect for this element. It can be positive or negative.

The mass defect is divided by the number of protons to obtain the mass change per proton. In the table and charts, this is "change / proton."

The mass defect is also divided by the number of neutrons to obtain the mass change per neutron. In the table and charts, this is "change / neutron."

There are 118 elements in the periodic table.
Above element 83, or Bismuth, all are radioactive.

Below 83, only element 43, or Technetium, and element 61, or Promethium, are radioactive.

Therefore there are 81 elements in this mass defect analysis.

2.2 Calculating Mass Change per Nucleons

Each atom has a number of these components:
1) protons,
2) electrons,
3) neutrons.

Item (1) is the element's atomic number.

The number in item (2) matches the number of protons, because the neutral atom has the same number of positive and negative charges.

Nucleons are protons and neutrons.

These names can be used:

NM = Nominal Mass of atom. This is the sum of protons and neutrons in the isotope that defines this element.

Therefore NM also represents the number of nucleons.

This value must be correct for this exercise.

Elements can have isotopes where the number of neutrons changes for a specific atomic number.

MM = Measured Mass of atom. This is the measured value of the specific isotope specified by NM. This value must have a suitable number of significant digits.

If the measured sample includes isotopes other than the one defined by NM, or if MM is rounded off, any difference from comparing masses is from that mistake, not from within the atom.

EMI = Expected Mass of Isotope, or its Nominal Mass.

NP = number Protons, or the atomic number.

NN = number Neutrons, from NM - NP

MCN = Mass change per Neutron, from a calculation
MCP = Mass change per Proton, from a calculation

The nominal mass of a neutron = $m_p + m_e$

EMI = (NP * mp) + (NN * ($m_p + m_e$))

The simpler calculation of the isotope's expected mass:

EMI = NM * (mp + me)

The calculation of MCN:

MCN = (MM − EMI) / NM

The calculation of MCP:

MCP = (MM − EMI) / NP

In simple terms, each atom has a difference in mass between measured and expected, or predicted, by the sum of its particles.

If the protons in the nucleus changed their apparent mass by MCP, then that assignment explains that mass defect.

If the neutrons in the nucleus changed their apparent mass by MCN, then that explains that mass defect.

Every atom, except for hydrogen, has more neutrons than protons in its nucleus.

As a result, MCN is lower in value than MCP.
The mass defect is more likely to be attributed to the neutrons, not the protons. This section concludes mass defect is a neutron behavior.

2.3 Calculating Neutron Mass

With a measured mass of a particular isotope, it is possible to calculate the mass of the neutrons by subtracting the mass of the protons and the orbiting electrons.

NM = Nominal Mass of the Isotope, or sum of its nucleons.

MM = Measured Mass of Isotope.

NP = number Protons, or the atomic number.

NN = number Neutrons, from NM − NP

The measured nominal mass of an atom is assumed to be this sum:

$$MM = (NP * (m_p + m_e)) + (NN * n^0)$$

Where m_e is the orbiting electron and n^0 is the neutron in the nucleus.

The average neutron mass can be calculated from:

$$n^0 = (MM - (NP * (m_p + m_e))) / NN$$

Average Neutron Mass will be used frequently so it will be ANM in this section.

2.3 Future New Formula for an expected atomic mass value

When an apparent mass loss becomes predictable for each element, then it can become part of the calculation,

MLN = Mass Loss per Neutron, to adjust an EMI.

The new calculation of the atom's expected mass is simple:

$$EMI = NP + (NN * (m_p + m_e + MLN))$$

This new equation accounts for the expected, apparent mass loss per nucleon in a particular element.

MLN seems to vary by element. The formula for this MLN behavior per element remains unknown, though it is apparently driven by the combination of NP with NN.

Some elements have a similar MLN, like nitrogen and oxygen.

Note:

The accepted mass of a neutron is never used for calculating the mass of an atom from its components because the accuracy of the neutron's measured mass is uncertain.

2.4 Mass Defect per element

The author compiled data from all elements to compare each for their measured value against the sum of their components. Most of the elements in this sample have a deficit or the measured is less than its sum.

A reference PDF is available with this spreadsheet of element data, for someone curious for a list of elements with their mass defect, whether positive or negative:

ZMass_Defect-DataN.pdf

Note: This PDF is formatted for 8.5x11" page and has 16 pages. Compressing that content into a smaller page is impractical.

One should note the radioactive elements usually lack a precise measured mass value.

Therefore, those few atoms cannot provide a measured mass defect.

There are several elements having its sum greater than measured. For consistent terms, these atoms have extra not missing mass.

2.4.N Neutron

A neutron is the combination of 2 particles, electron and proton.

Its expected mass = mp + me

Or 1.0078250322

Its measured mass, obtained from a deuterium analysis method is 1.0086649159

The difference is +0.0008398836

This is not a binding energy but is an increase in apparent mass.

This result suggests the measured mass of a neutron uses a wrong method.

This analysis continues with several simple isotopes. A neutron was also part of Section 1.

2.4.1.H Element 1 is Hydrogen

Hydrogen is the combination of a proton and electron, with the electron in orbit around the proton.
Its expected mass = $m_p + m_e$

= 1.0078250465 using current values
Or 1.0078250322 using new values (earlier in section 1)

Its measured mass is 1.00794

The published mass value does not have many digits after the decimal point suggesting it is rounded.

Rounded values cannot be used for a mass defect.

Hydrogen has this isotope distribution:

^{1}H 99.98%
^{2}H 0.02%

This distribution results in 1.00020, which does not match this published value.

Trying this isotope distribution:

^{1}H 99.206%
^{2}H 0.794%

This alternate distribution results in 1.00794, which matches the published mass value.

Some of these published values must be wrong.

In any case, the protium atom must be used not a value from an isotope mix of hydrogen.

2.4.1.1 Protium

Protium is the combination of a proton and electron, with the electron in orbit around the proton.

Its expected mass = $m_p + m_e$

Or 1.0078250322

Its measured mass is 1.0078250322

The difference is 0.

This is because the new masses for the electron and proton get this desired result.

With the original particle masses, the sum exceeded the measured. This difference must be explained by nuclear binding energy, but the proton is alone. This difference and sign clearly show the particle masses were too high.

The sum of the 2 current mass values is 1.0078250465

Protium should not have a mass defect when the nucleus is only a proton.

2.4.1.2 Element D or Deuterium or ^{2}H Isotope or H-2

Deuterium has 1 proton and 1 neutron in the nucleus.

Its measured mass is 2.0141017781

Its expected mass = m_p + + m_e + (m_p + m_e)

or = 2.0161986444

The difference is -0.0020968663
The average neutron mass is 1.0062767459

This ANM is only the neutron in this atom of only 3 components with the binding energy between neutron and proton.

The nominal mass of a neutron is the same as Protium: 1.0078250322

The difference is -0.0015482864

This is the mass loss between 1 proton and an electron in a neutron.

The average neutron mass can be calculated for any isotope having a valid measured mass.

2.4.1.3 Element T or Tritium or ^{3}H Isotope or H-3

Tritium has 1 proton and 2 neutrons in the nucleus.

^{3}H half-life is 12.32 years and decays to ^{3}He by β− decay

^{3}H measured mass 3.0160492

^{3}H expected mass $= m_p + m_e + (2 * (m_p + m_e))$

^{3}H expected mass is 3.0234750967

The difference is -0.0074258967

This is the apparent mass loss in bonds between 3 nucleons (1p + 2n).

The average neutron mass is 1.0041120839

This T ANM is less than that of Deuterium.

In the context of deuterium, this T ANM value relative to a normal neutron is the binding energy required for another proton-neutron bond after deuterium.

Adding a neutron to proton is like adding a neutron to a proton-neutron pair.

2.4.2.3 ^{3}He Isotope Tralphium

Tralphium has 2 protons and 1 neutron in the nucleus.

Its measured mass is 3.0160293

Its expected mass = $(2 * (m_p) + m_e + (m_p + m_e)$

Its expected mass is 3.0234750967

The difference is -0.0074457967

This is the apparent mass loss in bonds between 3 nucleons (2p + 1n).

The mass change per neutron is -0.0006705477

This value is less than that of tritium.

The average neutron mass is 1.0003792355

This ^{3}He ANM is less than that of Deuterium and Tritium.

In the context of deuterium, this ^{3}He ANM value relative to a normal neutron is the binding energy required for another proton bond after deuterium.

2.4.3 Comparing 3 Nucleons

The comparison between tritium and Tralphium could be crucial to understanding mass defect.

They are different combinations of protons and neutrons with a total of 3. The measured mass change is not from the number of particles but is from their types.
Comparing the measured values:

^{3}H measured mass 3.0160492
^{3}He measured mass is 3.0160293
^{3}H - ^{3}He = 0.0000199

The 1p + 2n atom is > 2p + 1n atom

Note:

After this analysis, the tiny mass defect is considered more likely driven by neutrons in the nucleus than protons.

Details are provided from published data but somehow This behavior must be tested and verified by experiments in a laboratory.

Comparing the neutrons:

The T average neutron mass is 1.0041120839

The ^{3}He average neutron mass is 1.0003792355
ANM from T - ^{3}He = +0.0037328484

The average of the 2 is more than the single when bonded with the opposite number or protons.

2.4.2.4 Element 2 is Helium ^{4}He

Helium ^{4}He has and 2 protons and 2 neutrons in the nucleus.

Its measured mass is 4.002602

Its expected mass 4.0313001290

The difference is -0.0286981290

This is the apparent mass loss in bonds between 4 nucleons (2p + 2n).

The mass change per neutron is -0.0071745322

Average Neutron Mass is 0.9934759678
This value is less than the others.

The difference is the "missing mass" claimed to be gone to nuclear binding energy.

Perhaps, mass was not converted into energy claimed to reside within the nucleus, where the energy cannot be measured as evidence.

The atom exhibits less mass. The old explanation is loss of mass. The new explanation is a nucleus has changed its reactivity to other masses, with the assumption this change occurs in one or more nucleons in the nucleus. There is no change in a physical quantity of mass.

The mass difference is divided by either the number of neutrons or protons to check the possible distribution of the particular mass defect within the atom's nucleus. In this book, this value is named Mass Change per Neutron, or just MCN, for a neutron or Mass Change per Proton, or just MCP.

The measured value can be used to calculate the Average Neutron Mass, or ANM, after subtracting the protons and orbiting electrons.

Helium exhibits less mass because the 4 particles in the nucleus are less reactive to other masses, so the result is it seems to have less mass.

2.5 Charts from Analysis

The first chart is presented before the explanations of some of the elements, so each element can be explained after first seeing how they compare.
First, here are all the elements from 1 to 92, charted with their respective mass change. Some have a positive change while most are negative.

The atomic number is below its value of change in mass per proton or neutron in this atom.

Elements 43 and 61, 84 to 89, and over 92, have a mass change value of zero because they are radioactive and are not measured accurately. 90 to 92 are radioactive but they have measured mass values having significant digits implying these 3 have accurate measurements.

Each extreme of a positive and negative value will be explained.

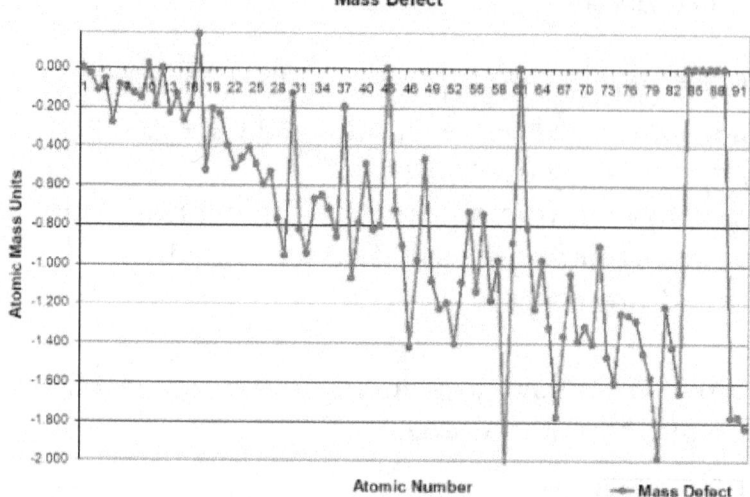

Mass Defect

Atomic Number

— Mass Defect

The elements having noticeable spikes will be explained.

Another chart will be presented after explaining which atoms have a large plus or minus value by a cause other than a mass defect.

This book proposes the measured mass defect is from the neutrons in the nucleus of the atom.

For each element, the measured mass subtracts the protons and orbiting electrons, and then is divided by the number of neutrons to get the average mass per neutron.

This value shows how much each neutron appears to change its mass.

As with the mass defect chart, there are a number of elements having an anomaly to be explained later.

For charting several of the radioactive elements, the measured value set to the expected value. With that match, the average mass per neutron is exactly one neutron.

Those points can serve as a reference to compare their values.

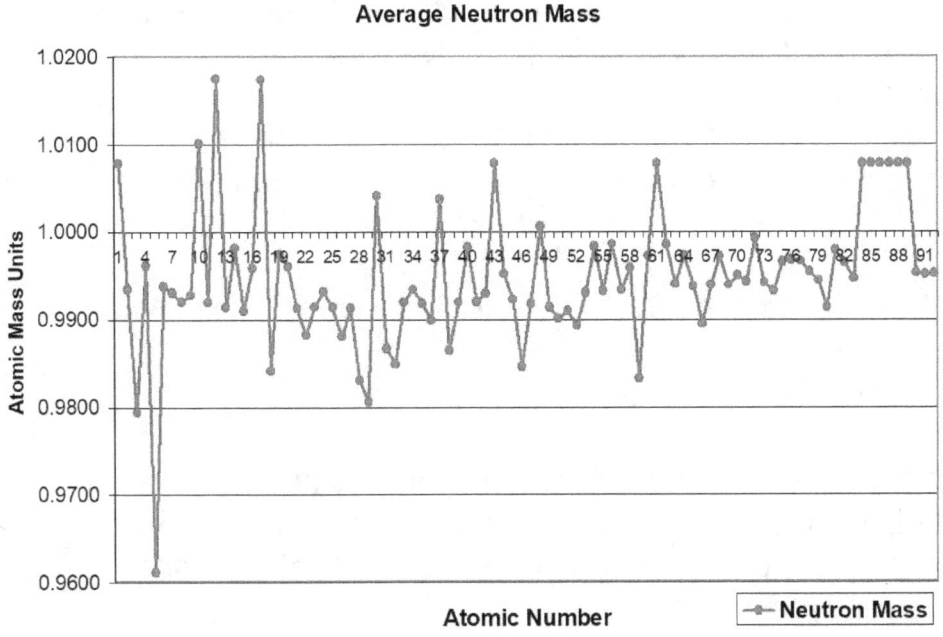

Other than the few anomalies, most elements have a very small change in apparent mass in each neutron.

A smaller set of elements, without the few anomalies, will be charted. This is without the outliers who are invalid for a mass defect check. Better scaling also enables a better perspective of the MCN values among the valid elements.

2.6 Initial Analysis

There several remarkable peaks in the chart for elements 1 to 92.

The difference in an atom's mass is calculated by subtracting the expected mass, the sum of its particles, from the measured mass. Each element has a defined count of protons and neutrons, enabling one to calculate its expected mass from its particles, electrons, protons, and neutrons.
The peaks are remarkable because their mass defect is increasing more or less than adjacent elements. When a defect is a negative value, it exhibits "missing mass."

When a defect is a positive value, it exhibits "extra mass."

Most have a negative value.

This implies these particular combinations of protons and neutrons have a noticeably different mass defect than atoms having a difference in nucleon count (protons plus neutrons) of only 1.

All the elements having a noticeable peak will be investigated.
This research is relevant to the topic of a neutron, not gravity, but this book has linked them.

The descriptions of an element never explicitly mention whether its atom has a measured mass defect.

At this time, a pattern of the mass loss in each element is not clear. But a tentative explanation is offered for a mass defect.

In some cases, the published atomic mass value appears to come from a different isotope, resulting in a large mass defect value with either sign, for missing mass or extra mass.
Here is a sample of mass defect values.

For each element:
Measured mass, Predicted mass, Difference, Mass Change per Proton or Neutron, Average Neutron Mass

Helium
Measured 4.002602
Predicted 4.0313001290
Difference -0.0286981290
MCP -0.0143490645
MCN -0.0071745322
ANM 0.9934759678

Lithium
Measured 6.941
Predicted 7.0547752257
Difference -0.1137752257
MCP -0.0379250752
MCN-0.0094812688
ANM 0.9793812258

Berylium
Measured 9.012183
Predicted 9.0704252902
Difference -0.0582422902
MCP -0.0145605725
MCN -0.0029121145
ANM 0.9961765742

Carbon
Measured 12.0096
Predicted 12.0939003869
Difference -0.0843003869
MCP -0.0140500645
MCN -0.0023416774
ANM 0.9937749678

Nitrogen
Measured 14.00643
Predicted 14.1095504514
Difference -0.1031204514
MCP -0.0147314931
MCN -0.0021044990
ANM 0.9930935392

Oxygen
Measured 15.99903
Predicted 16.1252005159
Difference -0.1261705159
MCP -0.0157713145
MCN -0.0019714143
ANM 0.9920537178

Neon
Measured 20.1797
Predicted 20.1565006448
Difference +0.0231993552
MCP +0.0023199355
MCN +0.0002319936
ANM +1.0101449678

Silicon
Measured 28.084
Predicted 28.2191009027
Difference -0.1351009027
MCP -0.0096500645

MCN -0.0006892903

ANM 0.9981749678
For reference:

The current measured mass of a neutron is 1.0086649159

Therefore, a change in mass per neutron of 0.01 for an atom is about a 1% loss in each of its neutrons.

2.7 Analysis of certain elements

There are several elements noticeable in the chart of the analysis results.

Elements 1 and 2 were detailed earlier.

2.7.3 Element 3 is Lithium.

Lithium has and 3 protons and 4 neutrons in the nucleus.

Its measured mass is 6.941

The lack of precision suggests this value is rounded. Most have at least 4-6 digits after the decimal.

Lithium has this isotope distribution:

^{6}Li 7.59%
^{7}Li 92.41%

This distribution results in 6.9241

With this change:

^{6}Li 5.9%
^{7}Li 94.1%

This distribution results in 6.941, which matches the published value.

The author must avoid invalid values. This value is invalid if it came from a sample having isotopes. The correct ^{7}Li is required.

Anyone wanting to understand mass defect must recognize correct values are crucial.

Lithium remains in the chart which presents the data available to the author.

Mass defect is valid only for a correctly measured value.

The expected mass is 7.0547752257

The difference is -0.1167752257

The mass change per proton is -0.0379250752

The mass change per neutron is -0.0094812688

Average Neutron Mass is 0.9793812258

The difference is the "missing mass" claimed to be gone to nuclear binding energy.

Perhaps, mass was not converted into energy claimed to reside within the nucleus, where it cannot be measured as evidence.

The atom exhibits less mass. The old explanation is loss of mass. The new explanation is a nucleus has changed reactivity to other masses, with the assumption this change occurs in one or more neutrons. There is no change in a physical quantity of mass.

Lithium exhibits less mass because the 7 particles in the nucleus are less reactive to other masses, so the result is it seems to have less mass.

There is some doubt whether the measured mass value came from a sample containing more than isotope.

2.7.4 Element 4 is Beryllium

.

Beryllium has and 4 protons and 5 neutrons in the nucleus.

Its measured mass is 9.012183

Its expected mass is 9.0704252902

The difference is -0.0582422902

The mass change per proton -0.0145605725

The mass change per neutron is -0.0029121145

Average Neutron Mass is 0.9961765742

The difference is the "missing mass" claimed to be gone to nuclear binding energy.

Perhaps, mass was not converted into energy claimed to reside within the nucleus, where it cannot be measured as evidence.

The atom exhibits less mass. The old explanation is loss of mass. The new explanation is a nucleus has changed reactivity to other masses, with the assumption this change occurs in one or more neutrons. There is no change in a physical quantity of mass.

2.7.5 Element 5 is Boron.

Boron has 5 protons and 6 neutrons in the nucleus.

Its measured mass is 10.806

The lack of precision suggests this value is rounded. Most elements have at least 4-6 digits after the decimal.

Its 2 isotopes have these abundances:

^{10}B 20%
^{11}B 80%

Observation:

The sum of this distribution is 10.80, which is not the published atomic weight of boron.

When using these defined values in place of the published percentage values:

^{10}B 19.4%
^{11}B 80.6%

The result is 10.806, which matches its published value.

The published percentages do not match the published result.

Correct mass values are crucial.

Boron remains in the chart which presents the data available to the author.

Its expected mass is 9.012183

The difference is -0.2800753547

The mass change per proton is -0.0560150709

The mass change per neutron is -0.0093358452

Average Neutron Mass is   0.9611458065
The ANM for boron is very low.

The more likely explanation is the published value was from the isotope mix which resulted in the published value.

2.7.6 Element 6 is Carbon.

Carbon has 6 protons and 6 neutrons in the nucleus.

Its measured mass is 12.0096

The lack of precision suggests this value is rounded. Most elements have at least 4-6 digits after the decimal.

Its 3 isotopes have these abundances:

^{12}C 98.9%
^{13}C 1.1%
^{14}C 1 part per trillion

Observation:

The sum of this distribution is 12.011, which is not the published atomic weight of carbon, but it is close.

When using these specific values in place of the published percentage values:

^{12}C 99.04%
^{13}C 0.96%

The result is 12.0096, which matches its published value.

The published percentages do not match the published result.

Correct mass values are crucial.

Carbon remains in the chart which presents the data available to the author.

Most elements have more than 4 digits after the decimal point unless the value was affected by the uncertainty with isotopes.

Carbon's 4 digits imply that possibility.

Its expected mass is 11.0860753547

The difference is -0.0843003869

The mass change per proton is -0.0140500645

The mass change per neutron is -0.0023416774

Average Neutron Mass is 0.9937749678
This ANM is low. If the measured mass was an isotope mix, then it is invalid.

2.7.7 Element 7 is Nitrogen.

Nitrogen has 7 protons and 7 neutrons in the nucleus.

Its measured mass is 14.00643

Its expected mass is 14.1095504514

The difference -0.1031204514

The mass change per proton is -0.0147314931

The mass change per neutron is -0.0021044990

Average Neutron Mass is 0.9930935392

2.7.8 Element 8 is Oxygen.

Carbon has 8 protons 8 neutrons in the nucleus.

Its measured mass is 15.99903

Its expected mass is 16.1252005159

The difference is -0.1261705159

The mass change per proton is -0.0157713145

The mass change per neutron is -0.0019714143

Average Neutron Mass is 0.9920537178

2.7.9 Element 9 is Fluorine.

Fluorine has 9 protons and 8 neutrons in the nucleus.

Its measured mass is 18.9984031630

The value has many significant digits.

But its expected mass is 17.1330255481

The difference is -0.1502724496

The mass change per proton is -0.0166969388

The mass change per neutron is -0.0016696939

Averageneutron mass is 0.9927977873
The explanation of the neutron decreasing its reactivity is a behavior change, not a decrease in a physical quantity called mass.

2.7.10 Element 10 is Neon.

Neon has 10 protons and 10 neutrons in the nucleus.

Its measured mass is 20.1797

The lack of precision suggests this value is rounded. Most elements have at over 4 digits after the decimal.

Its isotopes have these abundances:

^{20}Ne 90.48%
^{21}Ne 0.27%
^{22}Ne 9.25%

Observation:

The sum of this distribution is 20.1877, which is not the published atomic weight of neon.

When using these defined values in place of the published percentage values:

^{20}Ne 89.83%
^{21}Ne 2.37%
^{22}Ne 7.8%

The result is 20.1797, which matches its published value.

But its expected mass is 20.1565006448

The difference is +0.0231993552

This atom has a tiny amount of extra mass.

The mass change per proton is +0.0023199355

The mass change per neutron is +0.0002319936

This value is very close to zero, and it is positive.

Average Neutron Mass is +1.0101449678
If the mass value is from an isotope mix then the neon MCN is invalid.

2.7.12 Element 12 is Magnesium

Data are from Wikipedia.

Magnesium has and 12 protons and 12 neutrons in the nucleus.

Its measured mass is 24.304

The lack of precision suggests this value is rounded. Most elements have at least 4-6 digits after the decimal.

Its 3 isotopes have these abundances:

^{24}Mg 79%
^{25}Mg 10%
^{26}Mg 11%

Observation:

The sum of this distribution is 24.32, which is not the published atomic weight of magnesium.

When using these specific values in place of the published percentage values:

^{24}Mg 80%
^{25}Mg 9.6%
^{26}Mg 10.4%

The result is 24.304, which matches its published value.

The published percentages do not match the published result.

Correct mass values are crucial.

Magnesium remains in the chart which presents the data available to the author.

Mass defect is valid only for a correctly measured value.

The expected mass is 24.1878007738

The difference is +0.1161992262
The mass change per proton is + 0.0096832689

The mass change per neutron is + 0.0008069391

Average Neutron Mass is 1.0175083011

This element is unusual because the expected is less than measured. There is extra mass in this atom.

This is the opposite of binding energy.

The more likely explanation is the measured value was distorted by isotopes in the sample.

2.7.17 Element 17 is Chlorine.

Chlorine has 17 protons and 18 neutrons in the nucleus.

Its measured mass is 35.446

The lack of precision suggests this value is rounded. Most elements have at least 4-6 digits after the decimal.

Its isotopes have these abundances:

^{35}Cl 76%
^{37}Cl 24%

Observation:

The sum of this distribution is 35.48, which is not the published atomic weight of chlorine.

When using these specific values in place of the published percentage values:

^{35}Cl 77.7%
^{37}Cl 22.3%

The result is 35.446, which matches its published value.

The expected mass is 35.2738761284
The difference is +0.1721238716

The mass change per proton is + 0.0101249336
The mass change per neutron is + 0.0005624963

Average Neutron Mass is 1.0173874696

This element is unusual because the expected is less than measured. There is extra mass.

The more likely explanation is the measured value was distorted by isotopes in the sample.

The published isotope percentages do not match the published result. The sample had a different distribution.

2.18 Element 18 is Argon

Argon has 18 protons and 22 neutrons in the nucleus.

Argon's measured mass is 39.948
The lack of precision suggests this value is rounded. Most elements have more than 4 digits after the decimal.

Argon has this isotope distribution:

^{36}Ar 0.334%
^{38}Ar 0.063%
^{40}Ar 99.604%

This distribution results in 39.858, which is not the published value.
The percentages do not add up to 100%. There must be uneven rounding for the total at 104.56%

This comparison between 2 mass values must avoid invalid values. This value is invalid if it came from a sample having isotopes. The correct ^{40}Ar is required.

Correct values are crucial.

Argon remains in the chart which presents the data available to the author.

Mass defect is valid only for a correctly measured value for the expected nominal atomic weight.

Its expected mass is 40.3130012896

The difference is -0.5210012896

This loss is equivalent to almost 1/2 missing neutron in this nucleus.

The more likely explanation is the measured value is from a sample having an isotope mix, so the value cannot be used to check against the predicted value. The predicted value is based on a specific proton and neutron combination.

The mass change per proton is -0.0289445161

The mass change per neutron is -0.0013156598

Average Neutron Mass is 0.9841431554

2.7.28 Element 28 is Nickel
Nickel has 28 protons and 31 neutrons in the nucleus.

Nickel's measured mass is 58.6334
The lack of precision suggests this value is rounded. Most elements have more than 4 digits after the decimal.

Nickel has this isotope distribution:

^{58}Ni 68.077%
^{59}Ni trace
^{60}Ni 26.223%
^{61}Ni 1.140%
^{52}Ni 3.635%
^{64}Ni 0.926%

This distribution results in 58.6402, which is not the published value.
The percentages do not add up to 100%. There must be uneven rounding for the total at 99.801%

This comparison between 2 mass values must avoid invalid values. This value is invalid if it came from a sample having isotopes. The correct ^{59}N is required.

This is an unusual element because the isotope of its nominal atomic mass exists in only a trace amount.

Correct values are crucial.

Nickel remains in the chart which presents the data available to the author.

Mass defect is valid only for a correctly measured value for the expected nominal atomic weight.

Its expected mass is 59.4616769022

 The difference is -0.7682769022

The mass change per proton -0.0274384608

The mass change per neutron is -0.0008851116

Average Neutron Mass is 0.9830419064

The loss is almost equivalent to 1 missing neutron in this nucleus.

The more likely explanation is the measured value is from a sample having an isotope mix, so the value cannot be used to check against the predicted value. The predicted value is based on a specific proton and neutron combination.

2.7.29 Element 29 is Copper

Data are from Wikipedia.

Copper has and 29 protons and 35 neutrons in the nucleus.

Its measured mass is 63.546

The lack of precision suggests this value is rounded. Most elements have at least 4-6 digits after the decimal.

Its isotopes have these abundances:

^{63}Cu 69.15%

64Cu, ^{66}Cu, and ^{67}Cu are not found
^{65}Cu 30.85%

Observation:

The sum of this distribution is 63.6170, which is not the published atomic weight of copper.

When using these defined values in place of the published percentage values:

^{63}Cu 72.7%
^{65}Cu 27.3%

The result is 63.546, which matches its published value.

The published percentages do not match the published result.

When using these defined values in place of the published percentage values:

63Cu 74.7%
65Cu 25.3%

The result is 63.546, which matches the calculated value from summing its components.

The published percentages do not match the published result. The sample had a different distribution.

Correct values are crucial.

Copper remains in the chart which presents the data available to the author.

Mass defect is valid only for a correctly measured value.

The expected mass is 64.5008020634

The difference is -0.9548020634

The mass change per proton is -0.0329242091
The mass change per neutron is -0.0009406917

Average Neutron Mass is 0.9805449733

The loss is almost equivalent to 1 missing neutron in this nucleus.

The more likely explanation is the measured value is from a sample having an isotope mix, so the value cannot be used to check against the predicted value. The predicted value is based on a specific proton and neutron combination.

Copper remains in the chart which presents the data available to the author.

There is some doubt whether the measured mass value came from a sample containing more than isotope.

2.7.36 Element 30 is Zinc

Krypton has 36 protons and 48 neutrons in the nucleus.

Zinc 's measured mass is 65.38
The lack of precision suggests this value is rounded. Most elements have more than 4 digits after the decimal.

Krypton has this isotope distribution:

^{64}Zn 49.2%
^{66}Zn 27.7%
^{67}Zn 4.0%
^{68}Zn 8.5%
^{70}Zn 0.6%

This distribution results in 65.45, which is not the published value.
The percentages do not add up to 100%. There must be uneven rounding for the total at 100.01%

This comparison between 2 mass values requires correct values. This value is invalid if it came from a sample having isotopes. The correct ^{64}Zn is required.

Correct values are crucial.

Zinc remains in the chart which presents the data available to the author.

Mass defect is valid only for a correctly measured value for the expected nominal atomic weight.

Zinc's expected mass is 65.5086270957

 The difference is -0.1286270957

The mass change per proton -0.0042875699

The mass change per neutron is -0.0001225020

Average Neutron Mass is 1.0041499724

This loss is different than the trend in adjacent atomic numbers, implying zinc has a change from that trend.

The more likely explanation is the measured value is from a sample having an isotope mix, so the value cannot be used to check against the predicted value. The predicted value is based on a specific proton and neutron combination.

2.7.36 Element 36 is Krypton

Krypton has 36 protons and 48 neutrons in the nucleus.

Krypton's measured mass is 83.798
The lack of precision suggests this value is rounded. Most elements have more than 4 digits after the decimal.

Krypton has this isotope distribution:

^{78}Kr 0.36%
^{80}Kr 2.29%
^{82}Kr 11.59%
^{83}Kr 11.50%
^{84}Kr 56.99%
^{86}Kr 17.28%

This distribution results in 83.8940, which is not the published value.
The percentages do not add up to 100%. There must be uneven rounding for the total at 100.01%

This comparison between 2 mass values requires correct values. This value is invalid if it came from a sample having isotopes. The correct ^{84}Kr is required.

Correct values are crucial.

Krypton remains in the chart which presents the data available to the author.

Mass defect is valid only for a correctly measured value for the expected nominal atomic weight.

Krypton's expected mass is 84.6573027082

The difference is -0.8593027082

The mass change per proton is -0.0238695197

The mass change per neutron is -0.0004972817

Average Neutron Mass is 0.9899228925
This loss is equivalent to almost 1 missing neutron in this nucleus.

The more likely explanation is the measured value is from a sample having an isotope mix, so the value cannot be used to check against the predicted value. The predicted value is based on a specific proton and neutron combination.

2.7.37 Element 37 is Rubidium.

Rubidium has and 37 protons and 48 neutrons in the nucleus.

Its measured mass is 85.4678

The lack of precision suggests this value might be rounded. Many have at least 6 digits after the decimal.

Rubidium has this isotope distribution:

^{85}Rb 72.17%
^{87}Rb 27.83%

This distribution results in 85.5566, which does not match the published value.

With this change:

^{85}Rb 76.61%
^{87}Rb 23.39%

This distribution results in 85.4678, which matches the published value.

This suggests the possibility the measured mass came from a sample of isotopes.

Mass defect requires correct values. This value is invalid if it came from a sample having isotopes. The correct ^{85}Rb is required.

Rubidium's expected mass is 85.6651277405

The difference is -0.1973277405

The mass change per proton is -0.0053331822

The mass change per neutron is -0.0001111080

Average Neutron Mass is 1.0037140376

This loss is different than the trend in adjacent atomic numbers, implying rubidium has a change from that trend.

The more likely explanation is the measured value is from a sample having an isotope mix, so the value cannot be used to check against the predicted value. The rubidium MCN is in the chart, regardless of a concern with the precision of its mass value.

2.7.38 Element 38 is Strontium

Strontium has and 38 protons and 50 neutrons in the nucleus.

Its measured mass is 87.62

The lack of precision suggests this value is rounded. Many have at least 6 digits after the decimal.

Strontium has this isotope distribution:
^{84}Sr 0.56%
^{86}Sr 9.86%
^{87}Sr 7.0%
^{88}Sr 82.58%

This distribution results in 87.104, which does not match the published value.

The expected mass is 88.6886028372
 The difference is -1.0686028372
The mass change per proton is -0.0281211273
The mass change per neutron is -0.0005624225
Average Neutron Mass is 0.9864529755

The loss is equivalent to a missing neutron in this nucleus.

Mass defect requires correct values. This loss value is invalid if it came from a sample having isotopes. The correct ^{88}Sr is required.

Strontium remains in the chart which presents the data available to the author.

2.7.43 Element 43 is Technetium

Technetium has 43 protons and 55 neutrons in the nucleus.

Its measured mass is 97.90721

But its expected mass is 98.7668531596

The difference is -0.8596431596

Average Neutron Mass is 1.0078250322

The substantial difference is the missing mass" which is presumed lost to nuclear binding energy.

This loss is equivalent to almost 1 missing neutron in this nucleus.

Which isotope was measured is not certain.

This is the isotope distribution of element 43, where syn means it is synthesized and is not found in nature. The half-life of each isotope is also shown.

^{95}Tc syn, 8.61d

^{96}Tc syn, 4.3d

^{97m}Tc syn, 4.21 x 10^6 y

^{98}Tc syn, 91d

^{99}Tc trace, 2.111 x 10^5 y

^{98}Tc syn, 91d

^{99m}Tc syn, 8.61d

The element's nominal mass is 98, but its only isotope found in nature, in a trace amount, is 99, and is radioactive.
Technetium is unique for other reasons.

Excerpt from Wikipedia:

Technetium is a chemical element with the symbol Tc and atomic number 43. It is the lightest element whose isotopes are all radioactive, none of which is stable other than the fully ionized state of ^{97}Tc. Nearly all available technetium is produced as a synthetic element. Naturally occurring technetium is a spontaneous fission product in uranium ore and thorium ore, the most common source, or the product of neutron capture in molybdenum ores.

The silvery gray, crystalline transition metal lies between manganese and ruthenium in group 7 of the periodic table, and its chemical properties are intermediate between those of both adjacent elements.

(Excerpt end)

Observation:

This excerpt describes an incredible behavior.

The nucleus of ^{97}Tc is stable only when having no orbiting electrons.

This might be the only nucleus which depends on external electrons for its stability.

The measured mass deficit already suggests an unusual nucleus. Perhaps the mass deficit is affecting this atom in this way. The more likely explanation is the measured isotope was not ^{98}Tc.

2.7.44 Element 44 is Ruthenium

Ruthenium has 44 protons and 57 neutrons in the nucleus.

Ruthenium has 1 more proton and 2 more neutrons than Technetium, the preceding element.

Ruthenium's measured mass is 101.07

The lack of precision suggests this value is rounded. Most elements have at least 4-6 digits after the decimal.

Its expected mass is 101.7903282563

The difference is -0.9004883208

Ruthenium is missing mass almost equivalent to a neutron.

The mass change per proton -0.0163710967

The mass change per neutron is -0.0163710967

Average Neutron Mass is 0.9951876944

The substantial difference is the missing mass" which is presumed lost to nuclear binding energy.

The loss is equivalent to almost 1 missing neutron in this nucleus.

There is some doubt whether the measured mass value came from a sample containing more than the correct isotope.

Ruthenium has 7 stable isotopes. The more likely explanation for the significant mass deficit is the measured mass was not of ^{101}Ru.

2.7.59 Element 59 is Praseodymium

Praseodymium has 59 protons and 82 neutrons in the nucleus. It has no isotopes found in nature.

Praseodymium's measured mass is 140.90766

Its expected mass is 142.1033295460

The difference is -2.0125645460

The loss is equivalent to almost 2 missing neutrons in this nucleus.

The mass change per proton is -0.0341112635

The mass change per neutron is -0.0004159910

Average Neutron Mass is 0.9832815622

The difference is the missing mass" which is presumed lost to nuclear binding energy.

2.7.60 Element 60 is Neodymium

Neodymium has 60 protons and 84 neutrons in the nucleus.

Neodymium's measured mass is 144.42

The lack of precision suggests this value is rounded. Most elements have at least 4 digits after the decimal.

Neodymium has this isotope distribution:

^{142}Nd 27.2%
^{143}Nd 12.2%
^{144}Nd 23.8%
^{145}Nd 28.3%
^{146}Nd 17.2%
^{148}Nd 5.8%
^{150}Nd 5.6%

This distribution results in 144.3570, which is not the published value.
The percentages do not add up to 100%. There must be uneven rounding for the total at 100.1%

This comparison between 2 mass values must avoid invalid values. This value is invalid if it came from a sample having isotopes. The correct ^{144}Nd is required.

Anyone wanting to understand mass defect must recognize correct values are crucial.

Neodymium remains in the chart which presents the data available to the author.

Mass defect is valid only for a correctly measured value for the expected nominal atomic weight.

Its expected mass is 145.1268046427

The difference is -0.8868046427

The mass change per proton is -0.0147800774
The mass change per neutron is -0.0001759533

Average Neutron Mass is 0.9972678341
The loss is equivalent to almost 1 missing neutron in this nucleus.

The more likely explanation is the measured value is from a sample having an isotope mix, so the value cannot be used to check against the predicted value. The predicted value is based on a specific proton and neutron combination.

2.7.61 Element 61 is Promethium

Promethium has 61 protons and 84 neutrons in the nucleus.
Excerpt from Wikipedia:

Promethium is one of only two radioactive elements that are followed in the periodic table by elements with stable forms, the other being Technetium.

The element also has 18 nuclear isomers, with mass numbers of 133 to 142, 144, 148, 149, 152, and 154 (some mass numbers have more than one isomer). The most stable of them is promethium-148m, with a half-life of 43.1 days; this is longer than the half-lives of the ground states of all promethium isotopes, except for promethium-143 to 147. In fact, promethium-148m has a longer half-life than its ground state, promethium-148.

(Excerpt end)

Observation:

The connection between elements 43 and 44 being similar to the pair of 61 and 62 might be relevant to some scientists, but those connections are beyond the scope of this book.

2.7.62 Element 62 is Samarium

Samarium has 62 protons and 88 neutrons in the nucleus.

Samarium's measured mass is 150.36

The lack of precision suggests this value is rounded. Most elements have at least 4-6 digits after the decimal.

Samarium has 7 isotopes so this is probably from their distribution.

But its expected mass is 151.1737548362

The difference is -0.8137548361

The mass change per proton -0.0131250780

The mass change per neutron -0.0001491486

Average Neutron Mass is 0.9985778182

The substantial difference is the missing mass" which is presumed lost to nuclear binding energy.

The total difference is equivalent to almost 1 missing neutron in this nucleus.

There is some doubt whether the measured mass value came from a sample containing not only the correct isotope.

2.7.67 Element 67 is Holmium

Holmium has 67 protons and 98 neutrons in the nucleus.

Holmium's measured mass is 164.930328

But its expected mass is 166.2911303198

The difference is -1.3608103198

This loss is like a missing neutron in this nucleus.

The mass change per proton is -0.0203106018

The mass change per neutron is -0.0002072510

Average Neutron Mass is 0.9939392127

2.7.90 Element 90 is Thorium

Thorium has 90 protons and 142 neutrons in the nucleus.

Thorium's measured mass is 232.0377

But its expected mass is 233.8154074799

The difference is -1.7777074799

The loss is like a missing neutron in this nucleus.

Average Neutron Mass is 0.9939392127

2.7.91 Element 91 is Protactinium

Protactinium has 91 protons and 140 neutrons in the nucleus.

Protactinium's measured mass is 231.03588

Its expected mass is 232.8075824477

The difference is -1.7717024477

This loss is like a missing neutron in this nucleus.

Average Neutron Mass is 0.9951700148

2.7.92 Element 92 is Uranium

Uranium has 92 protons and 146 neutrons in the nucleus.

Uranium's measured mass is 164.930328

But its expected mass is 166.2911303198

The difference is -1.3608103198

The loss is like a missing neutron in this nucleus.

Average Neutron Mass is 0 0.9952671715

2.8 Mass Defect Summary

All of the elements which have the largest mass defect are also possibly using an isotope with a different number of neutrons, not the correct measured mass value for a correct calculation when seeking the nuclear binding energy.

The calculations were done even with the suspicious values which cause noticeable spikes in the charts.

These elements had a possible measurement from a sample of more than one isotope: 3, 5, 10, 12, 17, 18, 20, 28, 29, 30, 36, 37, 38, 43, 44, 59, 60, 62.

18 out of 81 is a substantial set lacking an accurate value.

Each had its loss somewhat different from the trend in adjacent elements.

Unfortunately for this analysis, many of the atomic mass values have less than 6 digits after the decimal point.

If any value has been rounded down, then that reduction cannot be called a mass defect.

Nearly all the elements have a small mass loss, apparently due to neutrons in the nucleus.

This finding suggests a new mechanism must be defined in particle physics for the context of an atom's nucleus. Clearly, the current atomic model is incapable of explaining the flexibility of a neutron's mass to match these measured values.

2.9 Mass Defect Mechanism

The author has proposed a new behavior for bonds between subatomic particles in the nucleus, where the mass reactivity in the neutrons can change. In most atoms its apparent mass decreases but in a few atoms, its neutrons seem to have more mass than they should. This new behavior must explain the observed difference between the measured and expected mass of many elements.

This behavior suggests the neutrons in a tight bond with protons can change their apparent mass, which is just a measure of their reactivity to other masses. This is observed even in the simple neutron itself, where its mass is more than the sum of electron and proton.

Nearly every element exhibits a difference between is expected mass and its measured mass.

The descriptions of a mass defect imply this difference in measured mass must be explained by energy equivalence to that change in mass

This behavior can be derived from a new mechanism for the force of gravity.

A neutron in a nucleus can have its reactivity to other masses changed, usually reduced.

The Standard Model has no flexibility in its explanation of a neutron consisting of only 3 quarks, where 3these 3 quasi-particles are never observed outside of particle accelerators.

It is already known the measured mass of a neutron is greater than its sum of an electron and proton. This increase has not been explained. However, the accuracy of this mass measurement was questioned in section 1.

When in a nucleus, the apparent mass of the neutron drops to below its sum.

All the atoms are created by fusion, or by radioactive decay. That means the protons and neutrons were literally smashed together, under extreme pressure to form the nucleus.

The potential impact exists where its reactivity to other masses gets damaged in the process, so its measured mass is reduced.

The nucleosynthesis sequence of creating the observed distribution of elements has known problems, like its unlikely sequence through unstable isotopes.

The non-fusion solar model proposes transmutation of elements on the photosphere. This process is also called cold fusion because extreme temperatures and pressures are not required.

The author suspects the process of building a nucleus against the strong repulsive force of the current nucleus somehow reduces the reactivity of the neutron to other masses, while in the nucleus.

2.10 Mass Defect Conclusion

Before making specific observations, the possible lack of published atomic mass values must be addressed.

Chlorine and magnesium have anomalies.

These atoms are prevalent and should enable learning the cause of their anomaly.

The neutron is a critical component in particle physics and right now its measured mass comes from an analysis of deuterium which has only a proton and neutron in its nucleus. Unfortunately, this nucleus has a known mass defect which is not truly understood. This draws into question the accuracy of a neutron's mass.

This proposed change in behavior within a nucleus could be controversial but this exercise explains the observational data.

The MCN can change for each element. Someone else can determine an algorithm for that behavior in a nucleus. This book is about gravity.

To prevent the reader having doubt about related topics, this book tried to cover other topics when slightly connected to the main topics.

It should be noted the author recommends a tiny change in the mass of a proton and an electron.

Currently, their values do not add up to the hydrogen atom's measured mass.

There is only a single proton in the nucleus, so claiming the proton exhibits an apparent loss of mass as result of nuclear binding energy is an awkward proposal.

Maintaining the verified ratio of proton mass to electron mass is the basis of this change. The only other value requiring verified precision is the measured mass value of the protium atom.

This tiny change to reduce the mass defect in the protium to zero means all elements will have a tiny reduction in their calculated mass defect.
Currently those particle masses are based on the carbon-12 atom, which exhibits nuclear binding energy.

Those particle masses should not result in the hydrogen atom having a mass defect.

That mass ratio was not changed to get the desired result.

The author assumes the verification of the ratio between electron and proton masses not know the experimental basis for its current value, which has 11 significant digits.

Excerpt:

The number enclosed in parentheses is the measurement uncertainty on the last two digits [11]. The value of μ is known to about 0.1 parts per billion.

(Excerpt end)

This proposed change affects the foundation of particle physics when recommending a change in the method of calculating the mass for an electron and proton.

Excerpt from Wikipedia:

"In 1803 John Dalton proposed to use the (still unknown) atomic mass of the lightest atom, that of hydrogen, as the natural unit of atomic mass. This was the basis of the atomic weight scale."

Later oxygen was used, and after its isotopes were discovered, carbon-12 became the new benchmark atom.

The bottom line for this decision on such a fundamental change in particle physics is whether all agree the expected mass of the hydrogen (protium) atom must equal the sum of a proton and electron.

This book has other recommendations for physics including a mechanism for the force of gravity.

3 New Atomic Model

3.1 Background

A new atomic model includes 2 important observations in this book about particle physics:

1) the mass of an electron and proton should be changed slightly.

The measured mass of the protium must match the sum of electron and proton.

It is impossible for a single proton to possess nuclear binding energy.

2) the neutron has a flexibility not explained in the current model.

The observed difference in the mass of an atom's nucleus, from the sum of its particles, requires an explanation.

The structured atomic model (SAM) matches most of the observational data of the elements.

The researchers of SAM claim the structure of the nucleons in the nucleus determine some of the properties of the element.

This book is not about SAM but SAM is probably part of the future of particle physics.

Here is an example of a nucleus having structure, from the SAM web site "etherealmatters.org"

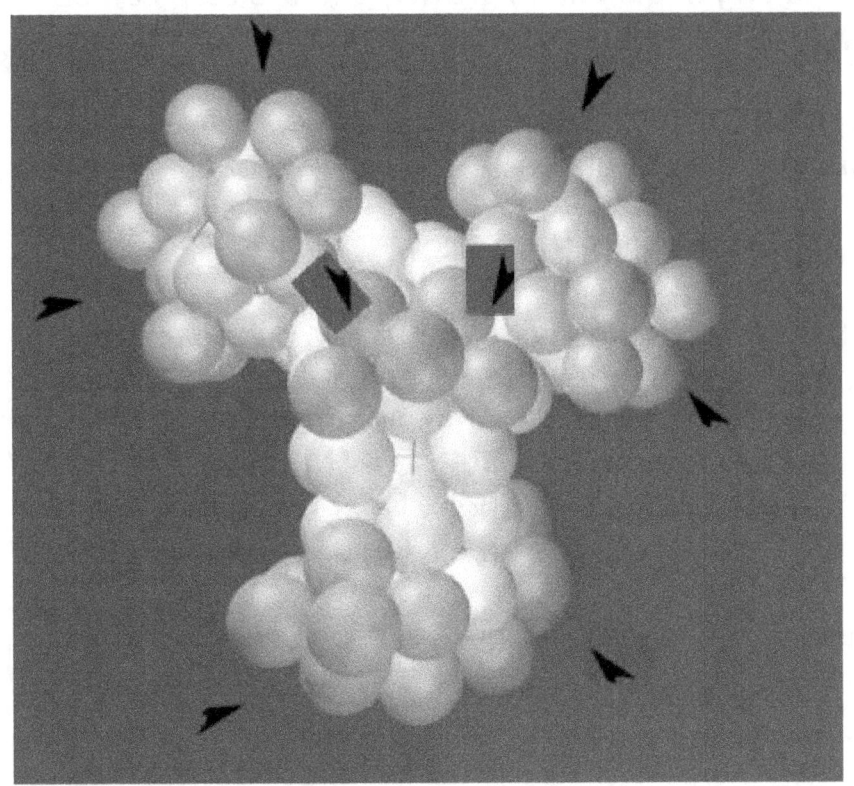

SAM is an important development in particle physics.

However, the public information about SAM does not mention mass defect at the time of writing this.

After completing this book on gravity, a future step can be notifying the SAM team about the mass defect behaviors.

All the behaviors driven by an atomic model is a separate topic from the force of gravity.

3.2 Comparison of the models

This author began this book by expanding on the concept of his simple particle model, which could build on SAM for its nuclei structures while SAM has no effect on the proposed mechanism for gravity.
The mass defect is a known anomaly for measuring the mass of an atom. It requires an explanation by any atomic model.
The results of an analysis of mass defect for each element are presented in section 2.

As a result, this new atomic model attempts to explain a mechanism for mass defect, by the particles the nucleus being affected by new nucleons added to the mix.

The observed mass defect values do not directly result in a specific mass loss in each particular bond, like proton-to-proton, and proton-to-neutron.

Mass defect has an elusive explanation.

In contrast, the standard model proposes nuclear binding energy. This calls a mass defect "missing mass" which is badly similar to the term "dark matter." The similarity in terms implies a mistake is being made here also.

Dark matter is invoked when magnetic fields are ignored.

Missing mass is invoked when a mechanism in the atom's nucleus is ignored.

The neutron is known to have a mass greater than the sum of electron and proton.

That should suggest the neutron is not as simple as the standard model expects.

Even if my new atomic model is not completely valid, at least it makes an effort at explaining a mass defect. The standard model claims it is just "missing" as if that is a suitable explanation.

The expectation is this book's analysis of the mass defect can lead to a valid explanation of the behavior, rather than tolerating atomic mass is just "missing." A common mechanism is implied when nitrogen and oxygen, adjacent elements in the periodic table, have similar values.

The efforts by the SAM team indicate the structure within the nucleus affects the elements chemical behavior. This is advancement for chemistry.

In stark contrast, the standard model offers nothing of substance beyond those interested in particle physics and its fragments.

The knowledge both a proton and neutron consist of 3 fragments contributed nothing to anyone outside that tiny group.

The standard model is essentially a game putting together a pattern of the fragments observing by breaking atoms.

That exercise has not explained the mechanism driving a mass deficit in a nucleus. That is probably because the behavior exists only in an intact nucleus.

I suggest the standard model and the LHC are no longer useful for the further advancement of science.

The human body is a complex system of many layers. There is no practical value in knowing whether:

a) protons have 3 pieces.

b) breaking atoms results in other pieces.

LHC is now just research only for the curious, but with no benefit for others. This is expensive research, serving a limited market.

Particle physics is in the domain of diminishing returns.

The opposite is cosmology where new instruments can offer new insights at scales from the solar system, to the extent of the observable universe which is constrained only by the technology to measure it.

An atom's measured mass usually does not exactly match the sum of its components.

Above the scale of atoms, the measured mass of many particles matches the sum of all the particles.
When in an atomic nucleus, this result is not always true.

This mismatch also applies to a neutron.

That should have triggered physicists to question all the behaviors of a neutron. This book took that step.

One could point out many have explained there is no particle called a photon. One such explanation was referenced in section 1, regarding the energy in light.

The photon is critical in the standard model to explain various behaviors, suggesting those explanations have no physical basis.

3.3 Building an Atom

Several atoms are described to help one begin to understand how a stable atom is built. This is an observation, not an explicit definition of a mechanism.

The author cannot explain the strong force, which holds nucleons together to maintain a stable nucleus.

The hydrogen isotopes were detailed in the Mass defect section, including their measured and predicted values. Those isotopes involve adding a number of neutrons to a single proton. This analysis continues with the isotopes of hydrogen, which has 1 proton, and helium, which has 2 protons. Each has isotopes with a number of neutrons.

First, terminology is defined; then the alpha particle is noted.

3.3.1 Nuclear Binding Energy
The binding energy maintains a nucleus.

Excerpt from Wikipedia:

Nuclear binding energy is the energy required to disassemble a nucleus into the free, unbound neutrons and protons it is composed of. It is the energy equivalent of the mass defect, the difference between the mass number of a nucleus and its measured mass. Nuclear binding energy derives from the nuclear force or residual strong force, which is mediated by three types of mesons.
(Excerpt end)

Observation:

Mass defect was explained in an earlier section.

3.3.2 Excess Energy

The excess energy is a comparison of one nucleus to that of Carbon-12.

Excerpt from Wikipedia:

The mass excess of a nuclide is the difference between its actual mass and its mass number in atomic mass units. It is one of the predominant methods for tabulating nuclear mass. The mass of an atomic nucleus is well approximated (less than 0.1% difference for most nuclides) by its mass number, which indicates that most of the mass of a nucleus arises from mass of its constituent protons and neutrons.

Thus, the mass excess is an expression of the nuclear binding energy, relative to the binding energy per nucleon of carbon-12 (which defines the atomic mass unit). If the mass excess is negative, the nucleus has more binding energy than 12C, and vice versa. If a nucleus has a large excess of mass compared to a nearby nuclear species, it can radioactively decay, releasing energy.

(Excerpt end)

Observation:

The mass defect analysis found no atom having excess energy in its nucleus. Initially, there were some but those were from the measured mass increased by the isotope mix, which is a mistake, not a mass defect.

3.3.3 Alpha particle

The alpha particle is an important part of an atom's decay, which can occur sometime after being built.

Excerpt from Wikipedia:

Alpha particles, also called alpha rays or alpha radiation, consist of two protons and two neutrons bound together into a particle identical to a helium-4 nucleus. They are generally produced in the process of alpha decay, but may also be produced in other ways. Alpha particles are named after the first letter in the Greek alphabet, α. The symbol for the alpha particle is α or α2+. Because they are identical to helium nuclei, they are also sometimes written as He^{2+}

or $^4_2He^{2+}$

indicating a helium ion with a +2 charge (missing its two electrons). If the ion gains electrons from its environment, the alpha particle becomes a normal (electrically neutral) helium atom ^4_2He.

Due to the mechanism of their production in standard alpha radioactive decay, alpha particles generally have a kinetic energy of about 5 MeV, and a velocity in the vicinity of 4% the speed of light.

(Excerpt end)

Alpha decay is mentioned. It is explained in the same Wikipedia topic:

The best-known source of alpha particles is alpha decay of heavier (> 106 u atomic weight) atoms. When an atom emits an alpha particle in alpha decay, the atom's mass number decreases by four due to the loss of the four nucleons in the alpha particle. The atomic number of the atom goes down by exactly two, as a result of the loss of two protons – the atom becomes a new element.

In contrast to beta decay, the fundamental interactions responsible for alpha decay are a balance between the electromagnetic force and nuclear force. Alpha decay results from the Coulomb repulsion between the alpha particle and the rest of the nucleus, which both have a positive electric charge, but which is kept in check by the nuclear force. In classical physics, alpha particles do not have enough energy to escape the potential well from the strong force inside the nucleus (this well involves escaping the strong force to go up one side of the well, which is followed by the electromagnetic force causing a repulsive push-off down the other side).

(Excerpt end)

Observation:

The "Coulomb repulsion" was mentioned but no value was provided.

Section 1 had calculations of several charged particle combinations. The electric force between 2 adjacent protons was calculated to be:

F_e = +81.7639597476 N

The "well" with the strong force must suppress ejections possible by this internal force.

An alpha particle has substantial kinetic energy on its ejection.

Since alpha decay occurs only with the most massive elements (having more protons), those nuclei have an even stronger force driving an alpha particle ejection.

3.3.4 Hydrogen and helium isotopes

Hydrogen and helium have isotopes which were detailed in the Mass Defect section.

Deuterium demonstrates 1 proton and 1 neutron can form a stable nucleus. This pair is not mutually repulsive.

A diproton, or just a proton pair, is unstable.

Tralphium demonstrates the combination of 2 protons and 1 neutron can form a stable nucleus. This proton pair is mutually repulsive but ^{3}He is stable.

For any combination to be stable, the creation of this nucleus with a neutron can overcome in force of repulsion between the protons. This combination of 3 creates the "well" required for the strong force.

Therefore, a neutron is required to accompany 1 or 2 protons for the strong force to exist to maintain a nucleus. As protons increase in number, neutrons generally increase in number faster. Lithium has 3 protons with 4 neutrons. The heaviest elements have many more neutrons than protons.

The mechanism of how neutrons hold protons together to create a "well" is not clear.

The strong force must be a behavior dependent on neutrons within a nucleus.

3.3.5 Summary of hydrogen and helium isotopes

The isotope and the number of digits in its mass value

10d = 10 digits after decimal point

^{2}H 10d, 1 proton, 1 neutron
^{2}H ANM 1.0062767459

^{3}H 7d, 1 proton, 2 neutrons
^{3}H ANM 1.0041120839

^{3}He 7d, 2 protons, 1 neutron
^{3}He ANM 1.0003792355
^{4}He 6d, 2 protons, 2 neutrons
^{4}He ANM 0.9934759678

Comparing ANM values:

^{3}H is less than ^{2}H,
^{3}He is less than ^{3}H, ^{4}He less than ^{3}He and ^{3}He more than ^{4}He.

Note:

Perhaps someone wishes to have mass values in MeV rather than amu, or atomic mass units.

The multiplier for amu conversion to MeV is
931.4941024200

For reference:

proton mass 1.0072764523

proton in MeV 938.2720748613

electron mass 0.0005485799

electron in MeV 0.5109989425

In this analysis, the comparison is between elements.

As long as everything uses the same units, the unit selection is not critical for this analysis. A PDF is provided with all the element data, in amu, used in this book.

3.4 New Model Conclusion

The structured atomic model (SAM) identifies structure within an atom's nucleus. This structure of the protons and neutrons into subsets enables justification for other behaviors, like chemical behaviors with the electron configuration driven by the nucleus. The details of SAM are still in development.

This author expects an explanation will be found for the neutron behaviors in a mass defect. Helium has the high MCN value, as in closer to zero. This behavior must be driven by the nucleus configuration for each element.

4 Mechanism for the force of Gravity

4.1 Background

Isaac Newton was quoted as saying:

"You sometimes speak of gravity as essential and inherent to matter. Pray do not ascribe that notion to me, for the cause of gravity is what I do not pretend to know, and therefore would take more time to consider of it."

Excerpts are from Wikipedia.

When the text might have an uncertain source, then Observation is used, but when inserted too frequently it is distracting.

Mathematical Principles of Natural Philosophy is a work in three books by Isaac Newton, in Latin, first published 1687.

Newton defined the behavior of F = ma.

A definition of mass:

Mass is both a property of a physical body and a measure of its resistance to acceleration (a change in its state of motion) when a net force is applied. An object's mass also determines the strength of its gravitational attraction to other bodies.

With the publication of "A Dynamical Theory of the Electromagnetic Field" in 1865, [James Clerk] Maxwell demonstrated that electric and magnetic fields travel through space as waves moving at the speed of light.

Observation:

Isaac Newton (178 years earlier) could not know how an electric field works so he could not propose a gravity field.

4.2 A New Mechanism for the Force of gravity.

This explanation begins with examining Maxwell's equations.

The video abut Maxwell's equations, noted in a previous section, offers a reader a visual presentation, which could supplement the text below about those equations.

There is at least one other theory of gravity proposing it is based on the electric force between charges in the subatomic particles in the respective masses.
Such a theory assumes there is only an electric force and gravity is just a manifestation of it, enabling a bipolar behavior.

This proposes gravity is a separate attractive force being dependent on the medium, just like the electric force.

Everyone knows the 2 inverse-square forces are similar in their equation format but there is a notable difference between them: gravity is much weaker.

When there are 2 similar mutual forces with similar behaviors but one is much weaker, then probably each force is uniquely affected by the medium.

The following explanation omits the calculus but hopefully this has enough detail.

Maxwell's equations define several properties of "free space" and those values define the rate of propagation of light through that free space.

Now, they can be considered properties of the medium, aka, the aether, which is whatever unknown "stuff" permeates the universe.

The medium defines the rate of propagation of the synchronized electric and magnetic fields within light.

Most know light travels slower through glass or water than through air or space.

The diffraction index is the factor defining the change in light velocity by the medium.

Essentially, the medium has a measurable resistance to the changing of electric and magnetic fields. During the propagation of light, both fields are oscillating or in continuous change.
Light is more complicated then that simple statement because different wave lengths have different behaviors like X-rays which can be either penetrating or shielded by different media. The color violet is slower than red through a glass prism.

At the foundation of Maxwell's equations are 2 constants which define how the medium affects changes in an electric field or a magnetic field:

the permittivity of free space, $\varepsilon 0$, epsilon-nought
the permeability of free space, μ, mu

These factors become Coulomb's constant.

The Electric force is described by Coulomb's law.

$$F = ke * (q1 * q2) / r^2$$

where ke is Coulomb's constant (ke $\approx 8.99 \times 10^9$ $N \cdot m^2 \cdot C^{-2}$), q1 and q2 are the signed magnitudes of the charges, and the scalar r is the distance between the charges. The force of the interaction between the charges is attractive if the charges have opposite signs (i.e., F is negative) and repulsive if like-signed (i.e., F is positive).

In very simple terms, there is a force between any 2 charges. This electric force is reduced by 2 factors:

1) ke from the medium,

2) r from the distance.

Observation:
The units of ke are essentially a ratio of force in an area relative to charge.

Free space defines a factor within ke resulting in a force reduction between charges.

After noting the role of free space in electromagnetism, the force of gravity is considered next.

The force of gravity is defined by Newton's Law of Universal Gravitation.

$$F = G * (m1 * m2) / r^2$$

where F is the gravitational force acting between two objects, m1 and m2 are the masses of the objects, r is the distance between the centers of their masses, and G is the gravitational constant.

The measured value of the gravitational constant is approximately $6.674 \times 10^{-11} \cdot m^3 / kg \cdot s^{-2}$

One might notice the mix of units in the force equation. There are seconds in its constant's units but there is no time variable in the factor value it multiplies, which has only these units: kg^2 / m^2

The units of force are $kg \cdot m \cdot s^{-2}$

This mismatch of time in units is reminiscent of Planck's equation, noted in section 1.

Wikipedia has a topic Cavendish Experiment describing how G was initially calculated using the oscillation of a torsion bar in an experiment taking 20 minutes.

Its current accepted value is by measurement, not by a calculation using defined "free space" parameters.

I assume the units of ke are just a remnant from measurements during experiments.
Dropping the s^{-2} units from the constant (again, there is no time value in the equation) leaves only:

m^3 /kg

This factor, which looks like inverse density, defines a ratio between:

a) a distance (though this m^3 in the numerator implies a distance is being treated as a volume) and

b) a participating mass, with kg in denominator.

This factor is essentially a ratio of a distance per mass.

The multiplication results in less force per kg, because the G value is much < 1

Instead of "Gravitational constant" this factor could be named "Gravitational Gradient" because the force reduces over distance, based on the medium, though the underlying parameters are not identified as Maxwell did for an electromagnetic constant.

This ratio might be considered a "free space" behavior for a gravity field.

Electric and magnetic fields required individual free space parameters. A gravity field apparently requires its own free space parameter.

4.3 Defining a New Mechanism for Force of Gravity

Newton did not propose a mechanism for mass to drive its force of gravity.

I propose the instantaneous force of gravity is the result of a mass field around every proton and electron.
Both particles already have an accepted electric field.

Both particles and atoms behave as expected for combinations of charges.

The 2 fields around both fundamental particles are different though the resulting mutual force affects both participants similarly.

The electric field is either attractive or repulsive while the gravity field is only attractive.

This simple assumption means all of Maxwell's equations for a static electric field and its mutual force also apply to this static gravity field and its mutual force.

Gravity is not electrical so permittivity for a capacitance in free space does not apply.

For gravity, open space is just a "resistance" to the force. That word is used in the definition of mass. The distributed free space resistance for this particular force explains why the force of gravity is so different between the masses of proton and electron compared to the force between their charges.

This new gravity field is NOT the accepted gravitational field around a sphere of uniform density causing free fall acceleration to smaller bodies near its surface.
Calling it a gravity field compared to an electric field is appropriate when one applies Maxwell's field equations.

The following is the Wikipedia description of an electric field but mixed with changes for an application to a gravity field highlighted.

I hope this from/to approach for changing a description from an electric field to a gravity field is clear. The exercise should reveal their similar behavior, to support this hypothesis.

1 From:

An electric field (sometimes E-field is the physical field that surrounds each electric charge and exerts force on all other charges in the field, either attracting or repelling them. Electric fields originate from electric charges, or from time-varying magnetic fields. Electric fields and magnetic fields are both manifestations of the electromagnetic force, one of the four fundamental forces (or interactions) of nature.

1 To:

A **gravity** field (sometimes **G-field** is the physical field that surrounds each **mass** and exerts force on all other **masses** in the field, attracting them. **Gravity** fields originate from **masses**. **Gravity** fields are manifestations of one of the four fundamental forces (or interactions) of nature.

G-field is used, not M-field, to avoid confusion with a magnetic field.

2 From:

The electric field is defined mathematically as a vector field that associates to each point in space the (electrostatic or Coulomb) force per unit of charge exerted on an infinitesimal positive test charge at rest at that point.The derived SI units for the electric field are volts per meter (V/m), exactly equivalent to newtons per coulomb (N/C).

2 To:

The **gravity** field is defined mathematically as a vector field that associates to each point in space the (force per unit of **kg** exerted on an infinitesimal positive test **mass** at rest at that point.The derived SI units for the **gravity** field are N per meter (**N**/m), exactly equivalent to newtons per kg (N/kg).

Image and caption (**3**) from Wikipedia Electric Field:

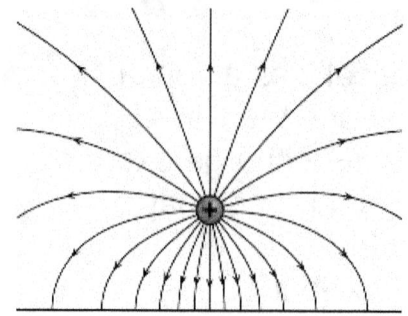

3 From:

Electric field of a positive point charge suspended over an infinite sheet of conducting material. The field is depicted by electric field lines, lines which follow the direction of the electric field in space.

3 To:

Gravity field of a point **mass** suspended over an infinite **span** of **gravity** conducting medium. The field is depicted by **gravity** field lines, lines which follow the direction of the **gravity** field in space.

4 From:

The electric field is defined at each point in space as the force (per unit charge) that would be experienced by a vanishingly small positive test charge if held at that point. As the electric field is defined in terms of force, and force is a vector (i.e. having both magnitude and direction), it follows that an electric field is a vector field. Vector fields of this form are sometimes referred to as force fields. The electric field acts between two charges similarly to the way the gravitational field acts between two masses, as they both obey an inverse-square law with distance.

4 To:

The **gravity** field is defined at each point in space as the force (per unit **mass**) that would be experienced by a vanishingly small positive test **mass** if held at that point. As the **gravity** field is defined in terms of force, and force is a vector (i.e. having both magnitude and direction), it follows that a **gravity** field is a vector field. Vector fields of this form are sometimes referred to as force fields. The **gravity** field acts between two **masses** similarly to the way the gravitational field acts between two masses, as they both obey an inverse-square law with distance. This is the basis for **Newton's** law, which states that, for stationary **masses**, the **gravity** field varies with the source **mass** and varies inversely with the square of the distance from the source. This means that if the source **mass** were doubled, the electric field would double, and if you move twice as far away from the source, the field at that point would be only one-quarter its original strength.

5 From:

This is the basis for Coulomb's law, which states that, for stationary charges, the electric field varies with the source charge and varies inversely with the square of the distance from the source. This means that if the source charge were doubled, the electric field would double, and if you move twice as far away from the source, the field at that point would be only one-quarter its original strength.

5 To:

This is the basis for **Newton's** law, which states that, for stationary **masses**, the **gravity** field varies with the source charge and varies inversely with the square of the distance from the source. This means that if the source **mass** were doubled, the **gravity** field would double, and if you move twice as far away from the source, the field at that point would be only one-quarter its original strength.

6 From:

The electric field can be visualized with a set
of lines whose direction at each point is the same as the
field's, a concept introduced by Michael Faraday, whose
term 'lines of force' is still sometimes used. This illustration
has the useful property that the field's strength is
proportional to the density of the lines. The field lines are
the paths that a point positive charge would follow as it is
forced to move within the field, similar to trajectories that
masses follow within a gravitational field.

6 To:

The **gravity** field can be visualized with a set
of lines whose direction at each point is the same as the
field's, a concept introduced by Michael Faraday, whose
term 'lines of force' is still sometimes used. This illustration
has the useful property that the field's strength is
proportional to the density of the lines. The field lines are
the paths that a point **mass** would follow as it is forced to
move within the field, similar to trajectories that masses
follow within a gravitational field. Field lines due to
stationary **masses** have several important properties,
including always originating from **point masses** and
terminating at other **masses**, they enter all **masses** at right
angles, and they never cross or close in on
themselves. The field lines are a representative concept;
the field actually permeates all the intervening space
between the lines. More or fewer lines may be drawn
depending on the precision to which it is desired to
represent the field.

7 From:

Field lines due to stationary charges have several important properties, including always originating from positive charges and terminating at negative charges, they enter all good conductors at right angles, and they never cross or close in on themselves. The field lines are a representative concept; the field actually permeates all the intervening space between the lines. More or fewer lines may be drawn depending on the precision to which it is desired to represent the field.

7 To:

Field lines due to stationary **masses** have several important properties, including always originating from **masses** and they never cross or close in on themselves. The field lines are a representative concept; the field actually permeates all the intervening space between the lines. More or fewer lines may be drawn depending on the precision to which it is desired to represent the field.

8 From:

$$E(x_0) = F / q_0 = q_1 / (X_1 - X_0)^2 \ldots$$

This is the electric field at point x_0 due to the point charge q_1; it is a vector-valued function equal to the Coulomb force per unit charge that a positive point charge would experience at the position x_0. Since this formula gives the electric field magnitude and direction at any point x_0 in space (except at the location of the charge itself, x_1, where it becomes infinite) it defines a vector field. From the above formula it can be seen that the electric field due to a point charge is everywhere directed away from the charge if it is positive, and toward the charge if it is negative, and its magnitude decreases with the inverse square of the distance from the charge.

The Coulomb force on a charge of magnitude q at any point in space is equal to the product of the charge and the electric field at that point

8 To:

$$G(x_0) = F / m_0 = m_1 / (X_1 - X_0)^2 \ldots$$

This is the **gravity** field at point x_0 due to the point **mass** m_1; it is a vector-valued function equal to the **gravity** force per unit mass that a point **mass** would experience at the position x_0. Since this formula gives the **gravity** field magnitude and direction at any point x_0 in space (except at the location of the **mass** itself, x_1, where it becomes infinite) it defines a vector field. From the above formula it can be seen that the **gravity** field due to a point **mass** is everywhere directed away from the mass, and its magnitude decreases with the inverse square of the distance from the **mass**.

The **gravity** force on a charge of magnitude q at any point in space is equal to the product of the **mass** and the electric field at that point

End of the from/to sequence of 8 steps.

Observation:

The Wikipedia descriptions of an electric field behavior frequently have a reference to a similar behavior in gravity.

In some cases, the "To" text needed no change because gravity was already there.

Excerpt from Wikipedia Electric Field:

Coulomb's law, which describes the interaction of electric charges: is similar to Newton's law of universal gravitation:

This suggests similarities between the electric field E and the gravitational field g, or their associated potentials. Mass is sometimes called "gravitational charge". Electrostatic and gravitational forces both are central, conservative and obey an inverse-square law.

Each charge field is diminishing with distance. Their mutual interaction results in a mutual force.

(Excerpt end)

Observation:

A gravity field from a mass behaves the same with another mass like a pair of charges. The difference is an electric field has polarity and interacts with only other electric fields, or with a magnetic field.

A gravity field interacts with only other gravity fields, and is not affected by an electric or magnetic field.

The force of gravity, between gravity fields of its participants being pervasive, is instantaneous and does not propagate.

Gravity field is also simpler. Changing electric and magnetic fields create the other. A changing a mass cannot create another type of field.

The universe has pervasive charge fields and gravity fields, with "lines" to describe their relative strength.

This theory does not change Newton's force of gravity equation, which has been verified numerous times. It only tries to explain its mechanism.

Relativity broke Newton's valid application of the force of gravity. Relativity must be dropped by physics because Newton's force remains valid. Wikipedia claims relativity superseded Newton's force, which is such an incredible mistake. Even more so, when one realizes relativity applies only to a special moving observer, becoming quite irrelevant to sciences like cosmology when there is no special observer.
Any unified field theory having no special observer must ignore relativity.

This rudimentary theory is offered just because it seems either relativity or dark matter are always involved in every discussion of gravity.
Both are invalid and must be ignored when redefining the force of gravity.

Gravity has complex behaviors like orbital resonances.

This book is seeking a mechanism, not defining the math for all the behaviors.

An alternative proposed by some individuals, where gravity has a bipolar behavior, or both attractive and repulsive, is more complex than a gravity field.

This proposed theory should be more productive than some other theories having an inconsistent behavior, because gravity has a consistent behavior through many experiments and accurate slingshot trajectories.

I suspect the free space parameter for gravity has a different origin. "free space" remains a mystery to physics.

I have an unrelated observation, which must be stated.

Free space parameters, like for the electric and magnetic fields, are defined by the universal medium, or the aether. It is impossible to know whether these values are consistent throughout the observable universe.

5 Newton's Gravity Cases

5.1 Newton's Laws of motion

Isaac Newton defined 3 laws of motion, which are not actually relevant to this book. They are included here to avoid confusing Newton's set of laws with Kepler's set of laws.

These laws from Wikipedia:

1)
In an inertial frame of reference, an object either remains at rest or continues to move at a constant velocity, unless acted upon by a force.

2)
In an inertial frame of reference, the vector sum of the forces F on an object is equal to the mass m of that object multiplied by the acceleration a of the object: $F = ma$. (It is assumed here that the mass m is constant.)

3)
When one body exerts a force on a second body, the second body simultaneously exerts a force equal in magnitude and opposite in direction on the first body.

Some also describe a fourth law which states that forces add up like vectors, that is, that forces obey the principle of superposition.

(Excerpt end)

5.2 Law of universal gravitation

Isaac Newton defined the law of universal gravitation, one of the main topics of this book.

Excerpt from Wikipedia:

Newton's law of universal gravitation is usually stated as that every particle attracts every other particle in the universe with a force that is directly proportional to the product of their masses and inversely proportional to the square of the distance between their centers. The publication of the theory has become known as the "first great unification", as it marked the unification of the previously described phenomena of gravity on Earth with known astronomical behaviors.
This is a general physical law derived from empirical observations by what Isaac Newton called inductive reasoning. It is a part of classical mechanics and was formulated in Newton's work Philosophiæ Naturalis Principia Mathematica ("the Principia"), first published on 5 July 1687. In today's language, the law states that every point mass attracts every other point mass by a force acting along the line intersecting the two points. The force is proportional to the product of the two masses, and inversely proportional to the square of the distance between them.

The equation for universal gravitation thus takes the form:

$$F = G * (m1 * m2) / r^2$$

where F is the gravitational force acting between two objects, m1 and m2 are the masses of the objects, r is the distance between the centers of their masses, and G is the gravitational constant.

(Excerpt end)

Observation:
That is Newton's force of gravity equation which has been used many times with success.

Wikipedia continued:

Newton's law has since been superseded by Albert Einstein's theory of general relativity, but it continues to be used as an excellent approximation of the effects of gravity in most applications. Relativity is required only when there is a need for extreme accuracy, or when dealing with very strong gravitational fields, such as those found near extremely massive and dense objects, or at small distances (such as Mercury's orbit around the Sun).

(Excerpt end)

Observation:

That mistaken claim of Newton being superseded by Einstein is one of the motivations for this book.

The claim relativity is required only for "extremely massive and dense objects" suggests a black hole which exists only in the theory of relativity and does not really exist. Newton's law cannot apply to fictional objects.

5.3 Free Fall

Excerpt from Wikipedia:

In Newtonian physics, free fall is any motion of
a body where gravity is the only force acting upon it. In the
context of general relativity, where gravitation is reduced to
a space-time curvature, a body in free fall has no force
acting on it.

In a roughly uniform gravitational field, in the absence of
any other forces, gravitation acts on each part of the body
roughly equally, which results in the sensation
of weightlessness, a condition that also occurs when the
gravitational field is weak (such as when far away from any
source of gravity).
(Excerpt end)

Observation:

Free fall seems to be a possible violation of Newton's force
of gravity.

If a person drops two objects having different masses, they
appear to fall with the same rate of acceleration. By
intuition, one could expect a heavier object would fall
faster.

This behavior has been demonstrated in a vacuum
chamber and on the surface of the Moon.

Rather than demonstrating a possible conflict, it
demonstrates the instantaneous behavior of gravity.

The experiment includes 3 masses, the two different objects, with an extremely much larger third object. In this experiment, the third is the Earth or Moon.
There is always the mutual force of gravity between the 3 objects.

A force is maintained on the two smaller objects to prevent their motion toward the larger one.

At the instant both are released, both accelerate at the same rate by the gravitational field of the much larger object.

The Wikipedia text for "free fall" lacks the description of the gravitational field behavior, which I expected to find in this topic, but Wikipedia has it in another.

Excerpt from Wikipedia for gravitational acceleration:

Using the integral form of Gauss' Law this formula [of Newton's force] can be extended to any pair of objects of which one is extremely more massive than the other — like a planet relative to any man-scale artefact. The distances between planets and between the planets and the Sun are (by many orders of magnitude) larger than the sizes of the sun and the planets. In consequence both the sun and the planets can be considered as point masses and the same formula applied to planetary motions. (As planets and natural satellites form pairs of comparable mass, the distance 'r' is measured from the common centers of mass of each pair rather than the direct total distance between planet centers.)

If one mass is much larger than the other, it is convenient to take it as observational reference and define it as source of a gravitational field of magnitude and orientation given by:

$$g = - (GM / r^2) * rv$$

Where M is the mass of the field source (larger), [r is the distance] and rv is a unit vector directed from the field source to the sample (smaller) mass. The negative sign just indicates that the force is attractive (points backward, toward the source).

Then the attraction force F vector onto a sample mass 'm' can be expressed as:

$$F = mg$$

Here g is the friction-less, free-fall acceleration sustained by the sampling mass 'm' under the attraction of the gravitational source. It is a vector oriented toward the field source, of magnitude measured in acceleration units. The gravitational acceleration vector depends only on how massive the field source 'M' is and on the distance 'r' to the sample mass 'm'. It does not depend on the magnitude of the small sample mass.

(Excerpt end)

Observation:

Just as the electric field used calculus, the gravitational field does as well.

5.4 Slingshot trajectory

When space probes follow a slingshot trajectory past a distant planet to increase the velocity of that probe, the validity of Newton's force equation is verified every time.

The site universetoday had a page titled:
How do gravitational slingshots work?

When NASA calculates a trajectory of a space probe using another body to change the probe's velocity, it uses the force of gravity defined by Newton. Curvature of space-time is not used.

The web page noted above has a description of how NASA calculates a slingshot to execute a change in a probe's trajectory; a video is provided also. NASA has certainly demonstrated their technique with numerous successful missions.

The calculation of a slingshot involves these critical values:
a) the mass of the probe
b) the mass of the planet
c) the velocity of the probe
d) the velocity of the planet.

During the probe's approach there is the mutual force of gravity between the two bodies where the paths of both bodies are affected simultaneously. Obviously the probe with a rather small mass is affected much more than the planet.

These calculations are based on the simple Newton force equation.

As described above, mismatch between two bodies results in free fall acceleration of the much smaller body toward the much larger body.

The complete algorithm for a trajectory must account for a possible transition from the mutual force affecting the probe's vector to a closer proximity changing to a free fall behavior.

With probe's rapid fly-by, it probably never changes its vector to be toward the heavier body.

Relativity is based on space-time curvature by a gravitational field. The special observer's path is assumed to curve by the other body's gravitational field.

It is impossible to know whether anyone attempted to use the tensor equations to verify the path being predicted matched the path predicted by Newton's gravity equations, one for the mutual force, and the other for possible free fall.
Curvature never involves the mass of the observer because the use of a gravitational field can ignore it, unlike the mutual force of gravity.

In relativity, though neither body is driven by a special observer, during a fly-by both bodies would be curving toward the other,
Curvature also never describes an affect on the body exerting its gravitational field which is affecting the special observer's path. Relativity is limited to only the special observer and their reference frame.

NASA never used relativity in its calculations for a slingshot trajectory. NASA does not use space-time curvature when a precisely calculated path is required.

Relativity assumed gravity had a velocity limit of c. NASA assumes gravity is instantaneous.

While not a disproof of relativity this application just shows relativity's space-time behavior would not enable a precise gravitational slingshot and was never used.

5.5 Orbital Resonances in Solar System

These resonances are difficult to explain with anything other than gravity.

Excerpt from Wikipedia:

In celestial mechanics, orbital resonance occurs when orbiting bodies exert regular, periodic gravitational influence on each other, usually because their orbital periods are related by a ratio of small integers. Most commonly this relationship is found for a pair of objects. The physical principle behind orbital resonance is similar in concept to pushing a child on a swing, where the orbit and the swing both have a natural frequency, and the other body doing the "pushing" will act in periodic repetition to have a cumulative effect on the motion. Orbital resonances greatly enhance the mutual gravitational influence of the bodies (i.e., their ability to alter or constrain each other's orbits). In most cases, this results in an unstable interaction, in which the bodies exchange momentum and shift orbits until the resonance no longer exists. Under some circumstances, a resonant system can be self-correcting and thus stable. Examples are the 1:2:4 resonance of Jupiter's moons Ganymede, Europa and Io, and the 2:3 resonance between Pluto and Neptune. Unstable resonances with Saturn's inner moons give rise to gaps in the rings of Saturn. The special case of 1:1 resonance between bodies with similar orbital radii causes large Solar System bodies to eject most other bodies sharing their orbits; this is part of the much more extensive process of clearing the neighbourhood, an effect that is used in the current definition of a planet.

A binary resonance ratio in this article should be interpreted as the ratio of number of orbits completed in the same time interval, rather than as the ratio of orbital periods, which would be the inverse ratio. Thus the 2:3 ratio above means Pluto completes two orbits in the time it takes Neptune to complete three. In the case of resonance relationships among three or more bodies, either type of ratio may be used (in such cases the smallest whole-integer ratio sequences are not necessarily reversals of each other) and the type of ratio will be specified.

(Excerpt end)

Observation:
Resonances noted were Neptune and Pluto, 3 large moons of Jupiter, and gaps in Saturn's rings.

Excerpt continued:

The orbits of Pluto and the plutinos are stable, despite crossing that of the much larger Neptune, because they are in a 2:3 resonance with it. The resonance ensures that, when they approach perihelion and Neptune's orbit, Neptune is consistently distant (averaging a quarter of its orbit away). Other (much more numerous) Neptune-crossing bodies that were not in resonance were ejected from that region by strong perturbations due to Neptune. There are also smaller but significant groups of resonant trans-Neptunian objects occupying the 1:1 (Neptune trojans), 3:5, 4:7, 1:2 (twotinos) and 2:5 resonances, among others, with respect to Neptune.

In the asteroid belt beyond 3.5 AU from the Sun, the 3:2, 4:3 and 1:1 resonances with Jupiter are populated by clumps of asteroids (the Hilda family, the few Thule asteroids, and the numerous Trojan asteroids, respectively).

(Excerpt end)

Observation:

There are interesting resonances with Neptune being "a quarter of its orbit away."

Excerpt continued:

In the asteroid belt within 3.5 AU from the Sun, the major mean-motion resonances with Jupiter are locations of gaps in the asteroid distribution, the Kirkwood gaps (most notably at the 4:1, 3:1, 5:2, 7:3 and 2:1 resonances). Asteroids have been ejected from these almost empty lanes by repeated perturbations. However, there are still populations of asteroids temporarily present in or near these resonances. For example, asteroids of the Alinda family are in or close to the 3:1 resonance, with their orbital eccentricity steadily increased by interactions with Jupiter until they eventually have a close encounter with an inner planet that ejects them from the resonance.
In the rings of Saturn, the Cassini Division is a gap between the inner B Ring and the outer A Ring that has been cleared by a 2:1 resonance with the moon Mimas. (More specifically, the site of the resonance is the Huygens Gap, which bounds the outer edge of the B Ring.)

In the rings of Saturn, the Encke and Keeler gaps within the A Ring are cleared by 1:1 resonances with the embedded moonlets Pan and Daphnis, respectively. The A Ring's outer edge is maintained by a destabilizing 7:6 resonance with the moon Janus.

(Excerpt end)

Observation:

Resonances cause gaps in the asteroid belt. A gap in Saturn's rings was mentioned above but here it is named.

Excerpt continued:

A Laplace resonance is a three-body resonance with a 1:2:4 orbital period ratio (equivalent to a 4:2:1 ratio of orbits). The term arose because Pierre-Simon Laplace discovered that such a resonance governed the motions of Jupiter's moons Io, Europa, and Ganymede. It is now also often applied to other 3-body resonances with the same ratios, such as that between the extrasolar planets Gliese 876 c, b, and e. Three-body resonances involving other simple integer ratios have been termed "Laplace-like"or "Laplace-type".

(Excerpt end)

Observation: An exoplanet system exhibits an orbital resonance.
Excerpt continued:

Several prominent examples of secular resonance involve Saturn.

A resonance between the precession of Saturn's rotational axis and that of Neptune's orbital axis (both of which have periods of about 1.87 million years) has been identified as the likely source of Saturn's large axial tilt (26.7°). Initially, Saturn probably had a tilt closer to that of Jupiter (3.1°). The gradual depletion of the Kuiper belt would have decreased the precession rate of Neptune's orbit; eventually, the frequencies matched, and Saturn's axial precession was captured into the spin-orbit resonance, leading to an increase in Saturn's obliquity. (The angular momentum of Neptune's orbit is 104 times that of Saturn's spin, and thus dominates the interaction.)

(Excerpt end)

Observation:

This is an interesting proposed relationship with distant Neptune affecting Saturn's axial tilt.

I have noted others have compared axial tilts between that of Saturn to Earth.

Excerpt from Wikipedia for Earth's rotation:

Earth's axial tilt is about 23.4°. It oscillates between 22.1° and 24.5° on a 41000-year cycle and is currently decreasing.

Observation:

Saturn's axial tilt is about 10% more than Earth's.

There is an online article about a 13:8 resonance for Venus and Earth, not mentioned above.

This page is online:

Venus; Its Curious Orbital Relationship to Earth

Excerpt:

The influence of Venus upon mankind and indeed upon the Earth is an interesting one. The Earth and Venus are locked in a curious orbital dance with one another that sees changes in the Earth's orbit over various cycles, most important of which are the 90,000 and 100,000-year cycles. These are built on the shorter cycle of eight years less two days. The Earth and Venus are engaged in a phase locking of their two orbits as described by chaotic dynamics. The two are orbiting in an almost exact 13:8 resonance, which has consequences for both planets. This phase locking has the consequence of driving the long ages of the Earth.

The Maya and Middle Eastern civilizations were aware of the cycles and both measured an eight-year cycle of Venus. The Maya incorporated the Venus cycle as one of the principle cycles of their calendar. They had an intense interest in Venus. During the course of their civilization, Venus was the harbinger of the annual rains and the marker for the planting of the maize crop. But, the cycle was slipping slowly out of alignment by almost two days over an eight-year period. Gradually, the marker no longer coincided with the advent of the rains. The priest-astronomers of the various temple complexes were aware of this and compensated for it, following by direct observation.

They certainly had the temple complexes lined up to the sky, They could help with other alignments when not part of their original design.

(Excerpt end)

Observation:

I thought there were other Earth-Venus resonances but a web search revealed no others.

5.5.1 LaGrange Points

LaGrange points are another type of orbital resonance.

Excerpt:

In celestial mechanics, the Lagrange
points (also Lagrangian points, L-points, or libration points)
are orbital points near two large co-orbiting bodies. At the
Lagrange points the gravitational forces of the two large
bodies cancel out in such a way that a small object placed
in orbit there is in equilibrium relative to the center of
mass of the large bodies.
There are five such points, labeled L1 to L5, all in the
orbital plane of the two large bodies. L1, L2, and L3 are on
the line through the centers of the two large bodies, while
L4 and L5 each act as the third vertex of an equilateral
triangle formed with the centers of the two large bodies.
L1, L2, L3 are unstable equilibria, whereas L4 and L5 are
stable, which implies that objects can orbit around them in
a rotating coordinate system tied to the two large bodies.
For each given combination of co-orbiting planetary bodies
there are five Lagrange points L1 to L5 for the Sun–Earth
system, and in a similar way there are
five different Lagrange points for the Earth–Moon system.
Several planets have trojan satellites near their L4 and
L5 points with respect to the Sun.

Jupiter has more than a million of these trojans.

Artificial satellites have been placed in orbits near to L1 and L2 with respect to the Sun and Earth, and with respect to the Earth and the Moon. The Lagrange points have been proposed for uses in space exploration.

(Excerpt end)

Observation:

LaGrange points are another orbital resonance which should be straightforward for the instantaneous force of gravity.
The equilibrium is awkward to explain with relativity where only a special observer has their path being curved.

5.6 Tides

Tides are a demonstration of the force of gravity.

5.6.1 Ocean tides

Excerpt, image, caption from Wikipedia:

Tides are the rise and fall of sea levels caused by the combined effects of the gravitational forces exerted by the Moon and the Sun, and the rotation of the Earth.

(Excerpt end)

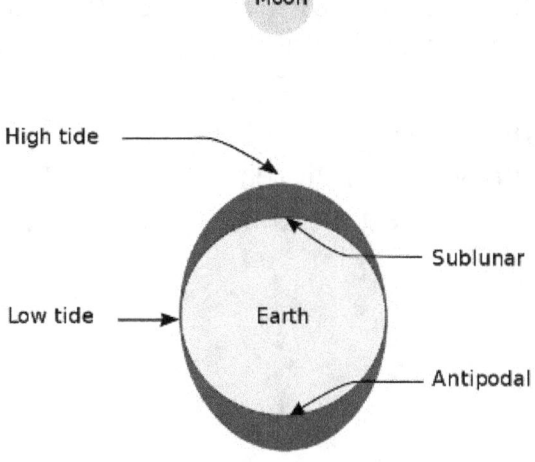

Caption:

Simplified schematic of only the lunar portion of Earth's tides, showing (exaggerated) high tides at the sublunar point and its antipode for the hypothetical case of an ocean of constant depth without land. Solar tides not shown.

5.6.2 earth tides

Excerpt from Wikipedia:

Earth tide (also known as solid Earth tide, crustal tide, body tide, bodily tide or land tide) is the displacement of the solid earth's surface caused by the gravity of the Moon and Sun. Its main component has meter-level amplitude at periods of about 12 hours and longer. The largest body tide constituents are semi-diurnal, but there are also significant diurnal, semi-annual, and fortnightly contributions. Though the gravitational forcing causing earth tides and ocean tides is the same, the responses are quite different.

(Excerpt end)

Observation:

Both an ocean tide and an earth tide is a distributed force of gravity, over a wide span of either water or land.

Relativity is for only an observer whose path is being affected by a gravitational field.

That scenario does not exist with tides. The tides behave consistently, though relativity cannot explain this behavior. Relativity involves an observer having their path diverted. Tides involve fluid behaviors and those behaviors involve more than gravity.

Everything in the universe is in motion by forces. None of them have a predetermined path. That is the requirement for a path to be affected by a gravitational field as determined by Einstein's equations.

Tides are a good example of why relativity should be dropped by physics. There is clearly an external force involved, driven between masses.

There are many efforts, like that by LIGO, or the 1919 solar eclipse, or the image of the M87 core, to find some way to confirm relativity but in every case there is another explanation available.
Gravitational waves are explained later. Earth tides are crucial to LIGO's claims of their wave detections having a very remote origin.

6 Kepler's Laws

Excerpt and the image from Wikipedia:

First law

The orbit of every planet is an ellipse with the Sun at one of the two foci.

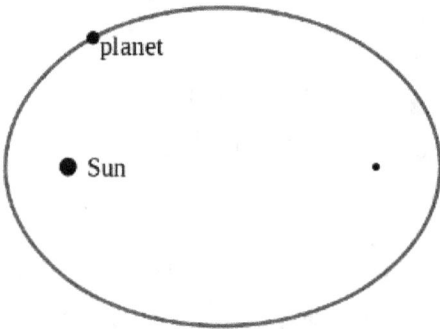

Second law

A line joining a planet and the Sun sweeps out equal areas during equal intervals of time.

The orbital radius and angular velocity of the planet in the elliptical orbit will vary.

Third law

The ratio of the square of an object's orbital period with the cube of the semi-major axis of its orbit is the same for all objects orbiting the same primary.
This captures the relationship between the distance of planets from the Sun, and their orbital periods.

Kepler enunciated in 1619 this third law in a laborious attempt to determine what he viewed as the "music of the spheres" according to precise laws, and express it in terms of musical notation. So it was known as the harmonic law.

(Excerpt end)

6.1 Ellipses

Kepler described the orbits with the Sun at a focus of the ellipse. That assumption is a mistake.
That focus is the instantaneous center of gravity in the solar system.

6.2 New Kepler's Third Law

Excerpt from Wikipedia:

Third Law: The Square of the orbital period of a planet is directly proportional to the cube of the semi-major axis of its orbit.

6.2.1 Its Equation

This law can be expressed by the formula: $T^2 = R^3$ where T is the time or period in years and R is the radius in AU. For examples:

a) Mercury T=0.2409 year and R= 0.3871 AU
b) Venus T=0.6152 year and R= 0.5352 AU
c) Earth T=1 year and R= 1 AU

These values conform to the equation.
The formula can also be represented also as a a ratio, like:
$X = T^2 / R^3$

so X should equal 1.0
if X is >1.0 then T is too high for R, so they are wrong for a valid ellipse. In other words, the pair of values is not the correct proportion.
Because published values are usually specified (or estimated) with a certain number of significant digits, the range of X is often from 0.998 to 1.002, or the comparison of the two values (from squared and cubed) is very close to equality.
All the planets in the solar system conform to this equation.

6.2.2 Its Problem with units

The units in this equation should not be $year^2 = AU^3$ because mixing these inconsistent units is invalid.
The current description lacks an explanation for the unit conflict.

The values in the equation are actually ratios with respect to Earth, so as a ratio of each value has no units. The description omits this crucial detail. The equation simply demonstrates the orbits in the system are proportional to each other by using ratios.

6.2.3 New Equation

This new simple equation has unit-less values from ratios using the ratio of the planet's orbit to a baseline orbit. The baseline could be Earth or another because all have same elliptical orbit relationship.

The equation uses these values:

BT = Baseline Time
BX = Baseline Axis

OT = Object Time
OX = Object Axis
TF = Time Factor
XF = Axis Factor

TF = OT / BT
XF = OX / BX

The new equation can be expressed with slightly different formats:

With no intermediate values:

$$(OT / BT)^2 = (OX / BX)^3$$

With the intermediate values:

$$TF^2 = XF^3$$

TF is the time scalar factor relative to the baseline time, or orbital period,
with the ratio of two values in whatever units are used to describe the period.

The values from each planet could be years, days, hours, or anything consistent. This is a ratio where its common units are critical.
The ratios require unit consistency but, as a result, the equation uses values with no units.

XF is the axis scalar factor relative to the baseline axis using a ratio in whatever units are used to describe the radius. The values from each planet could be AU, km, or anything consistent. This factor is from a ratio.
Currently these factors are the ratio for each planet's orbit to Earth's orbit as the baseline.
That is why the numbers of years and AU work. This equation can also use ratio values from a different planet as the basis than Earth.

The simple change to the third law: Instead of entering a planet's orbit values, their proportional values are entered.

6.2.4 Confirming the solution
All of the planets could have their factor based on their ratio to another planet because all their ellipses are proportionally consistent. This equation is based on the ratios between the respective elliptical orbits.

Each factor is a ratio from a body's value divided by a baseline value.
Currently we use Earth for the baseline for planets simply for its convenience, and because Kepler did. There is no mention whether considered systems of moons. If he had, this law would be different.

For Mercury based on Earth, TF= 0.3871 yr / 1 yr or 0.3871 and XF = 0.3871 (from 0.3871 AU / 1 AU).

6.2.5 Applying the solution to moons

Nearly every planet in our solar system has one or more moons, as well as the dwarf planet Pluto. Only Mercury and Venus have none.

As an initial test, Jupiter's set of moons can be verified with this new equation.

Jupiter has a number of large moons: 1 Io, 2 Europa, 3 Ganymede, 4 Callisto, 5 Amalthea, 6 Himalia, 7 Elara

If the moons other than Europa have their TF and XF based on the orbit of Europa, then this squared=cubed equation applies to all these moons.

Europa can have its baseline on Io.

Therefore, the new description applies to Jupiter's moons. This description change allows verification of orbits of moons as well as planets.

6.2.6 Applying this solution

The orbit parameters were entered into an Excel spreadsheet for all bodies in the solar system which could be found having this data (axis radius and time per orbit). This included all planets, all their moons and most identified asteroids.

Apparently only about 7000 of the (estimated) billions of asteroids have been identified but not all have their data published in a public archive like Wikipedia.

All the planets and asteroids in orbit around the Sun and all the moons were checked. The spreadsheet is a useful reference.

Earth has only 1 Moon but its orbit is roughly proportional to the International Space Station in its varying orbit (with distinct apogee and perigee).
Mars has 2 moons. Each can be the baseline for the other; both conform.
Jupiter has over 50 moons. Wikipedia notes some moons get lost and found again; 2 were found in 2004 and 2007 but both (unnamed) conform.

Europa can serve as the baseline for most of the moons. Only a few were not closely proportional to Europa. Carme has one of the longest
periods among Jupiter's moons. Harpalike provides a baseline. Both Megaclite and Chaldene have high eccentricities and Ananke provides a baseline.
The numbering of Saturn's moons is different in Britannica or Wikipedia.
I used the Britannica numbering because it was consistent with other references.

References are provided for:

a) Brittanica list of Jupiter's moons
b) Wikipedia list of Saturn's moons

In the case of Saturn's moons, the orbit of Janus can be the baseline for most moons. Janus can use Mimas for its baseline.
An issue with Saturn's moon collection is after the main moons, there are more moons with higher eccentricities.

Wikipedia describes it this way: Saturn has 24 regular satellites in prograde orbits not inclined. The remaining 58 are irregular satellites with a mix of prograde and retrograde.

Observation:

An unnamed moon found in 2004 is not closely proportional to Janus. This one is proportional to another unnamed one found in 2004.
In the case of moons around Uranus, the orbit of Miranda can be the baseline for most moons. Miranda can use Ophelia for its baseline.
Uranus also has irregular satellites. Only 3 long period moons, Caliban, Sycorax, and Margaret are not closely proportional to Miranda. these 3 are proportional to similar long period moons like Trincolo.

In the case of moons around Neptune, the orbit of Naiad can be the baseline for most moons. Naiad can use Thalassa for its baseline.

Only 2 high eccentricity moons Nereid and Psamathe are not closely proportional to Naiad. Both can use Halimede, which also has high eccentricity, for their baseline.

In the case of 5 moons around Pluto, the orbit of Styx can be the baseline for Charon. Styx can use Charon for its baseline. Nix, Kerberos and Hydra are more closely proportional to any in this set of 3 than to Styx.
2 of the high eccentricity moons (Nereid and Psamathe) are not closely proportional to Styx. Nereid is closely proportional to Halimede, while Psamthe is closely proportional to Nereid.

Apparently there are about 50 Nereids in a group around Pluto but only 3 (these 2 and Neso) have a name and data.

reference is available for this spreadsheet of orbit data, for the new 3rd law:

ZSolarSystem Data-N3.pdf

Note: This PDF is formatted for 8.5x11" page and has 15 pages. Compressing that content into a smaller page is impractical.

6.2.7 Exoplanets

Using new technology, in recent decades astronomers were able to detect planets in orbit around distant stars; these are called exoplanets.

The orbit parameters were entered into an Excel spreadsheet for many of the systems with several exoplanets which had the required data (axis and time per orbit).

Exoplanet Data was updated with check of Kepler's new 3rd law in this file:

ZExoplanet-Data-N3.pdf

The orbit data is in units ranging from AU, year, day. With the 3rd law equation based on ratios not on only the measured values using different units, the individual orbits are confirmed to conform to the baseline for the groups of exoplanets around their distant star.

Note: This PDF is formatted for 8.5x11" page and has 5 pages. Compressing that content into a smaller page is impractical.

6.2.8 New Third Law

Its description should change to something like this:

New Third Law:

Its Basis:
The baseline ellipse will be different for each primary, unless by coincidence two have the same mass.

Each body has its orbit stated as:

a) a unit-less value from a ratio of its period to a baseline period.

b) a unit-less value from a ratio from its axis to a baseline axis.

Its Description:

The square of the ratio from its orbital period is directly proportional to the cube of the ratio from its semi-major axis.

The square of the ratio from a body's orbital period is directly proportional to the cube of the ratio from a body's semi-major axis.

6.2.8.1 Basis for this new law.

There is a simple explanation for a baseline ellipse for other ellipses within each particular system, whether planets around a star, or moons around a planet.

The very large body in the middle of a collection of smaller bodies is the scenario of gravitational acceleration described earlier.
That suggests all the smaller bodies are in free fall acceleration toward the much heavier body.

However the final orbit has the system's center of gravity at the focus of the ellipse. This center of gravity must stabilize the orbit distribution. In our solar system, the giant planets have a wider separation between their orbits, while the smaller inner planets have less separation, while the asteroids in the main belt have much less separation.

The moons around Pluto have more than one baseline.

The most distant are similar while another set is similar.

Pluto is unique because its primary is essentially a binary. Pluto and Charon have similar masses and rotate about each other. This system lacks a single primary larger than all others.

6.2.9 Tidal Locked Moons

Nearly all moons in the solar system are in an orbit called a tidal lock or in synchronous rotation. Either means the moon maintains one side facing the primary while moving in its elliptical orbit around that primary.

None of the planets exhibit this behavior in their orbit around the Sun except Mercury, which has this resonance noted in Wikipedia:

Mercury rotates in a way that is unique in the Solar System. It is tidally locked with the Sun in a 3:2 spin–orbit resonance, meaning that relative to the fixed stars, it rotates on its axis exactly three times for every two revolutions it makes around the Sun.

(Excerpt end)

Definition from Wikipedia:

Tidal locking (also called gravitational locking, captured rotation and spin–orbit locking), in the best-known case, occurs when an orbiting astronomical body always has the same face toward the object it is orbiting. This is known as synchronous rotation: the tidally locked body takes just as long to rotate around its own axis as it does to revolve around its partner. For example, the same side of the Moon always faces the Earth, although there is some variability because the Moon's orbit is not perfectly circular. Usually, only the satellite is tidally locked to the larger body. However, if both the difference in mass between the two bodies and the distance between them are relatively small, each may be tidally locked to the other; this is the case for Pluto and Charon.

The effect arises between two bodies when their gravitational interaction slows a body's rotation until it becomes tidally locked. Over many millions of years, the interaction forces changes to their orbits and rotation rates as a result of energy exchange and heat dissipation. When one of the bodies reaches a state where there is no longer any net change in its rotation rate over the course of a complete orbit, it is said to be tidally locked. The object tends to stay in this state when leaving it would require adding energy back into the system. The object's orbit may migrate over time so as to undo the tidal lock, for example, if a giant planet perturbs the object.

(Excerpt end)

Observation:

The explanation says "may migrate over time."

When looking at the solar system now (which is certainly after some "time"), one might wonder which moons are NOT in synchronous rotation?

The answer is a very short list (data from Wikipedia):

Around Saturn:
Hyperion rotates 13d for 21.3d orbit

Ymir rotates 12h for 3.6y orbit

"Ymir is the largest retrograde irregular moon of Saturn" at about 18 km or 11 mi.

The outer irregular satellites [of Saturn] follow moderately to highly eccentric orbits, and none are expected to rotate synchronously as all the inner moons of Saturn do (except for Hyperion).

(Excerpt end)

Around Neptune:

Nereid - rotates 12h for 360d orbit

Around Pluto:

Styx - rotates 3.24d for 20.16d orbit
Nix - non-spherical rotates 44h for 24.9d orbit

Observation:

Cosmologists have an interesting caveat about a tidal lock changing over time.

When looking now, there are only 5 candidates. The current count of moons is either 175 or over 200 depending on the reference.

This research began expecting to find more than 5 and one of those is explicitly not a sphere so it could be excused. Perhaps I missed others.

After reading the description of this behavior, apparently the number should be more but is not. At only 2.5%, this behavior is more stable than suggested by the description.

Probably, Hyperion had to be explained in some way, like this, and the statement excused other moons to be found later.

The most striking statement, regardless of the % value is this:

"for example, if a giant planet perturbs the object."

Cosmologists explicitly accept Jupiter and Saturn roamed the solar system. Their excursion is placed in the Late Heavy Bombardment Period.

The mutual force of gravity causing a moon's orbit around the center of gravity in the planet's system is sufficient to maintain stability for a long time, until "perturbed" by another body.

7 Basics of Relativity

Einstein's Theory of Relativity is background independent meaning it never accesses an external coordinate. Whatever algorithms it uses cannot access a position in physical space. This is like taking a picture of where you are at your current time. No other observer can use that unique picture.

Einstein's Theory of Special Relativity defined a special observer as one who is moving through a gravitational field, but we are not that special observer when here on Earth.
Space-time is the name assigned to this moving observer's reference frame. As noted above, when being background independent, this space-time is unique to each observer.

The first book had a thorough description of the space-time context in relativity, but this short version should be sufficient here.

Einstein's Theory of General Relativity just "generalized" special relativity.

The Space-time section has more details about this special observer.

A black hole must be mentioned in this "basics" section because it is an entity only in relativity.

A claim that Newton's gravity cannot explain a black hole is just ridiculous.

Space-time curvature used tensor equations to manipulate the special observer's path.

With a large enough mass, the path would become a point in the space-time reference frame.

This path being stuck on a point in a reference frame is meaningless. This point is called a gravitational singularity, usually shortened to just singularity.

Physicists proposed an impossible result.

This mass in the real world must be present in this point in the reference frame. The reference frame is not a real thing which can hold an amount of matter. It has no size. The result is an impossible infinite density.

Physicists compounded the nonsense by claiming light cannot pass out of this point in this reference frame, leading to more nonsense like an event horizon.

Light is not affected by any reference frame, nor is it affected by gravity. Gravitational lensing is covered in a later section.

Physicists continued to compound the nonsense by claiming an accretion disk could form around this black hole.

This is totally ridiculous because the black hole existed only in a reference frame, while the matter remained unaffected in the real world.

Therefore, all observers can observe the matter at this black hole in space while the matter is supposedly also at a coordinate in the special observer's reference frame.

A black hole exists only for that unique moving observer.

Black holes are described further in a later section.

This initial description highlights the problem when physicists fail to recognize space-time is the special observer's reference frame and whatever happens within that context has no affect on the real world, and no other observer can use another observer's reference frame.

Each observer's reference frame is based on their current position in space and whatever they use for their time. All of that is their "reference."
It is impossible for two observers to occupy the same position in physical space (unless one is an apparition.

4.2 Gravitational Collapse

Excerpt from Wikipedia:

Gravitational collapse is the contraction of an astronomical object due to the influence of its own gravity, which tends to draw matter inward toward the centre of gravity. Gravitational collapse is a fundamental mechanism for structure formation in the universe. Over time an initial, relatively smooth distribution of matter will collapse to form pockets of higher density, typically creating a hierarchy of condensed structures such as clusters of galaxies, stellar groups, stars and planets.
A star is born through the gradual gravitational collapse of a cloud of interstellar matter. The compression caused by the collapse raises the temperature until thermonuclear fusion occurs at the center of the star, at which point the collapse gradually comes to a halt as the outward thermal pressure balances the gravitational forces. The star then exists in a state of dynamic equilibrium. Once all its energy sources are exhausted, a star will again collapse until it reaches a new equilibrium state.

(Excerpt end)

Observation:

This scenario is a clear violation of thermodynamics, several times.

A gas cannot self-compress to reduce its volume. It cannot decrease its volume, by itself, to increase pressure because pressure requires a surface. No surface exists in a gas cloud.

An external energy source is required to increase thermal energy in a system. There is no external energy source in this scenario.
This book is about gravity, not stars, but references are provided.

Stars are not powered by fusion in a core having an impossibly high pressure and temperature. This core supposedly sustains fusion which is radiating internal electromagnetic radiation whose "light pressure" is preventing catastrophic gravitational collapse. That light pressure requires a surface but there is none in a gas. The light we see is supposedly the light from fusion taking many years to propagate though solar layers of gas, until it escapes. The fusion probably generates gamma rays but that is not the observed spectrum. Instead the star has an external electric current to heat the solid core, whose thermal energy is transferred by convection to the photosphere where energy is released as thermal radiation.

All stars have a spectrum exhibiting thermal radiation. Stars are classified by their temperature obtained from their spectrum.

This mechanism for a star was described in the author's second book.

Pierre-Marie Robitaille has many videos about the failure of the fusion model for a star. One example is titled:

The Sun is NOT a Gaseous Plasma! The LMH Solar Model!

Gravitational collapse to initiate fusion violates thermodynamics.

It is also proposed for creating black holes and neutron stars.

These two entities are impossible by physics, regardless of their impossible creation mechanism.

Both are described in later sections.

For those wondering where elements are created by a star, if not in the core by fusion, they are created by transmutation, also known as cold fusion, or a low energy nuclear reaction, LENR. They are created on the solar surface, its photosphere.

The SAFIRE project duplicated the solar environment in a laboratory and confirmed the formation of elements not originally in the chamber.

More information can be found with:

Safire project was also noted earlier.

8 Space-time

Some theoretical physicists repeat the saying, "Spacetime tells matter how to move; matter tells spacetime how to curve."

This quote is credited to John Archibald Wheeler.

Excerpt from Wikipedia:

"John Archibald Wheeler was an American theoretical physicist largely responsible for reviving interest in general relativity in the United States after World War II."

(Excerpt end)

That task simply means he was presenting an "interesting'" version for the public, not one directly from Einstein, who never said that quote.

A coordinate system should be referenced to the physical space by the observer and only that observer selects their coordinate system.

Theoretical physicists can accept this laughable statement only by misunderstanding physics, relativity and a coordinate system.

One learns in a physics class about the important forces of gravity, electric, and magnetic, and the result of acceleration from a force on a mass. Objects move by the external forces acting on them.

Objects in the universe do not move using specific coordinates defined by an observer somewhere in the universe.

Excerpts from Wikipedia:

Postulates of special relativity
1. First postulate (principle of relativity)

The laws of physics take the same form in all inertial frames of reference.

General relativity generalizes special relativity and refines Newton's law of universal gravitation, providing a unified description of gravity as a geometric property of space and time, or space-time. In particular, the curvature of space-time is directly related to the energy and momentum of whatever matter and radiation are present. The relation is specified by the Einstein field equations, a system of partial differential equations.

(Excerpt end)

Observations:

Relativity is a background independent theory meaning it never uses physical coordinates. All its field equations are confined to the moving observer's reference frame. Relativity by its very design never affects any physical entity, only the special observer's reference frame whose dimensions can be curved in this mathematical exercise.

When cosmology tries to apply general relativity, the context for space-time is still confined to the special observer's reference frame. This reference frame uses no coordinates in physical space.

Nothing in the universe is a special observer with a non-inertial reference frame using commanded motion
Nothing in the universe moves using a coordinate system.

Everything in the universe moves as a result of forces acting on it. A change in kinetic energy requires a transfer of energy,
A coordinate system is not a source of energy.

The bending of light by space-time has been proposed. However, the propagation of the synchronized electric and magnetic fields is affected only by the medium, as defined by its diffraction index.
Light will never follow a path defined by a coordinate system.

The use of a coordinate system requires an intelligent observer to command a move described by a coordinate system. To suggest a planet or star will perform a commanded motion is simply foolish.

Some theoretical physicists appear detached from reality. That quote is an emphasis of a manipulated coordinate system lacking a connection to physical space. Space-time requires a person, a "special" observer moving through a gravitational field, to use their reference frame for curvature.

We are here on Earth, not on every celestial object to command their motions individually.

Classical physics was grounded in physical space with an established time increment for precise measurements while obtaining valid evidence from experiments to test and verify a theory.

We can measure celestial bodies in motion using our selected coordinate system. They move as affected by external forces.
They do not move according to someone's coordinate system.

Mankind has been observing the universe for a very long time without those bodies following a coordinate system.

7.1 Space-time reference frame in Graphics

Graphical representations of space-time curvature are indicating objects are assumed to me as directed by a coordinate system.

Image and caption from Wikipedia:

Two-dimensional analogy of spacetime distortion generated by the mass of an object. Matter changes the geometry of spacetime, this (curved) geometry being interpreted as gravity. White lines do not represent the curvature of space but instead represent the coordinate system imposed on the curved spacetime, which would be rectilinear in a flat spacetime.

Observation:

Only the special moving observer has their path distorted in relativity. This special observer must move to coordinates for the path to be distorted.

When we launch a satellite into orbit, like the one in the image, its trajectory follows that expected by the force of gravity from Earth affecting the probe's trajectory which began in a particular direction. Its path was NOT defined by a coordinate in a particular coordinate system. Once it is on its way, it is NOT changing its path by a not rectilinear space-time. Its path will curve by Earth's gravity. All objects in the universe respond to an external force. We are told to believe the probe is following spacetime but nothing really does that. Only forces affect motion. There are 3 fundamental forces in physics. To suggest anything else affects motion is beyond physics.

7.2 Space-time in Graphics

Graphical representations of space-time curvature are an intentional deception.

These images do not represent how the universe is observed when on or near the Earth.
We are not a special observer whose path is curved by that remote object's gravitational field. We see everything using our senses and instruments, not hidden within curved space-time.

This unedited image from NASA will help explain this deception.

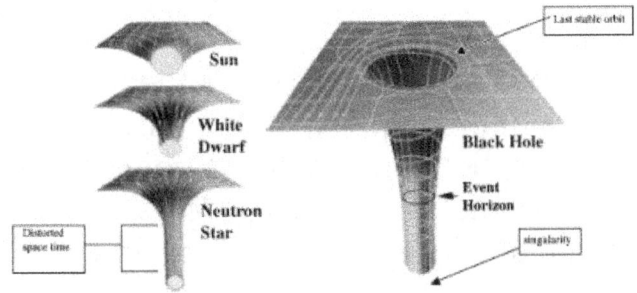

In relativity, when the observer is moving near an object with a gravitational field their 4-dimensional coordinate system will be curved so straight lines in Euclidean geometry are no longer straight. If the user defines their motion using coordinates in the distorted dimensions then their path will not be straight.

This curvature affects only the moving observer's coordinate system but no one else is affected.

Einstein's first postulate is "The laws of physics take the same form in all inertial frames of reference."

The left column in the image illustrates how the observer's space-time is curved when the observer is passing by the Sun, a white dwarf, or a neutron star.

For all other observers the Sun, the white dwarf, or the neutron star will be observed using classical physics, such as electromagnetic radiation.

The image is deceptive because there is no distinction between the observer moving past these objects and all other observers.

One could present an edited image to represent the view for all other observers by simply removing those curved graphics for the observer's space-time. At the lower left is the legend "distorted space time" explicitly noting the specific context for them in this image. That edited image removes the deception by showing the real universe, which all observers can observe and measure, and which is not affected by one observer's motion past a particular body in physical space.

The right column in the image has the most blatant deception.

The single arrow pointing to "Singularity" is actually pointing to 2 entities.

1) The physical mass at that location in physical space,

This mass is not shown here though each mass was shown in the left column.
The image could be edited as suggested to remove the graphs from the respective columns; then the mass should be shown here, consistent with the others, to help fix the deception for all observers other than the one moving (i.,e., non-inertial).

2) A point in the observer's reference frame or coordinate system.

The point is not in the image simply because a point has no size.

In basic geometry, the intersection of 2 lines is a point. The point is a specific coordinate in the coordinate system; a simple example of a point in 3-D is X1,Y2, Z3.

In the mathematical exercise of space-time curvature for an extreme mass, all the lines of the respective dimensions cross at a point called the singularity.

This singularity is called a black hole though technically it is a black point. There is no hole in anything; it is just a point in a coordinate system, the intersection of lines, defined by Euclid.

The deceptive graphic hides this disaster for physics with two simultaneous conflicting entities where one entity is a concept, just a point in a coordinate system, while the other is a physical mass.

For all other observers the mass is present and can be observed and measured and as a mass it is still subject to the force of gravity from other bodies.

Physicists chose to combine these two conflicting entities, resulting in something physically impossible.

The singularity is claimed to retain the mass and its gravitational field. However, this point has no size so the result is a gravitational field coming from a mass having infinite density.

There should be another arrow in the image next to that of Singularity and pointing to the same point but with the legend "Impossible"

There is no such thing as a black hole. This will be explained later in section 10.

Probably, if graphical representations of space-time curvature were not deceptive then impossible entities like black holes would go away.

Also, the mistaken claim of remote gravitational lensing should also go away having no justification for a remote curvature.

To present the reality of a proposed black hole, the image for most observers (except for the special observer) who have no distorted space-time, the bottom right should have this note inserted using the Sun's graphic icon (instead of O):

Note:
Milky Way SMBH has O x 4.1 million visible to all other observers.

(end of note)

That simple change to the figure clearly unveils the deception because there is NO huge real mass of that size, , observable by any observer other than the special observer, at that location claimed for that super massive black hole. Astronomers just claim this mass is present with no observational evidence. Everyone should "see" it, but go not.

The black hole is something we are told is there but we cannot observe it. That restriction makes the concept immediately suspicious.

9 Gravitational Wave

We observe a gravitational wave (GW) indirectly, only by its effect on the Earth's surface.

Excerpt from Wikipedia on LIGO:

The Laser Interferometer Gravitational-Wave Observatory (LIGO) is a large-scale physics experiment and observatory to detect cosmic gravitational waves and to develop gravitational-wave observations as an astronomical tool. Two large observatories were built in the United States with the aim of detecting gravitational waves by laser interferometry. These observatories use mirrors spaced four kilometers apart which are capable of detecting a change of less than one ten-thousandth the charge diameter of a proton.
The initial LIGO observatories were funded by the National Science Foundation (NSF) and were conceived, built and are operated by Caltech and MIT.

(Excerpt end)

Observation:

September 14, 2015 was the date of a critical failure for physics.

On that date, the LIGO staff turned on their system expecting to get a gravitational wave (GW) at some unknown time in the future.

They had no idea about anything of their completely untested system or even how often they would get events. They never used a real event to check whether their system properly analyzed that known event as a system verification.
If the known event resulted in wrong conclusions then they must fix it for their known test conditions.

After a very short time since initiating the system's executing its analysis, their system generated a wave form viewable by the staff. They claim they thoroughly verified all was well with their complex system and then eagerly announced this historic event.

YouTube has several videos featuring LIGO personnel describing their experience and surprise with the first GW.

There is no record of how they verified there was no terrestrial source.

That date was the date of a perigee, the event which occurs once a month or so, when the moon is closest to Earth in its orbit. Whether they checked this or the other lunar events is unknown.

This increased stress of Earth's crust was sufficient for the LIGO software to analyze it. LIGO cannot detect a theoretical GW directly so templates were developed on how such an event in space-time might possibly affect the Earth.

The only criteria were this signal detected from the most sensitive laser interferometers ever built must pass through the complex signal analysis.

There was no way to filter out a perigee because there is no record of testing with any terrestrial sources. The system was based on only templates for 4 event types and software. LIGO had the mistaken certainty only a remote GW could affect Earth's surface on such a global scale.

More mistakes followed by LIGO:
New Moon on October 12, 2015 triggered a never confirmed GW.
Full Moon on December 25, 2015 triggered a never confirmed GW.
Perihelion on January 4, 2017 triggered a never confirmed GW.

In the following years, these 4 types of terrestrial events resulted in many LIGO claims of GW detections never having an independent confirmation of any detail of their detailed descriptions.

All the LIGO claims were within a few days of these terrestrial events so there were no strays to suggest another source triggering their system.

By early 2019 I recognized this terrestrial source. Some Facebook posts on this LIGO mistake were met with derision in two groups. On November 10, 2019 I posted a comment to the LIGO Facebook page, open to public comments but not posts, predicting GW detections in 3 specific 5-day spans in that November. I could not specify exact dates because this LIGO system is unreliable even reporting 2 GW on the same day. However, over half the time LIGO reports a GW within only 2 days of the terrestrial source. All predictions were confirmed by LIGO GW claims.

I tried to stop the LIGO lunacy by providing these test results to the National Science Foundation, the primary resource provider for LIGO, in December 2019. LIGO's claims are all without merit. In early January 2020, NSF acknowledged my email but remained confident of LIGO so there was no public revelation of LIGO making its unverified claims.

On some subsequent LIGO facebook posts about specific historic GW events, I put my public comments to their public posts. LIGO never responded in any way to any of my contacts.

As a result of no revelation, the science of physics is still being taught the false claim of many theories being confirmed by LIGO.
Among the crucial theories NOT confirmed are: black holes, neutron stars, gravitational waves, and relativity in general.

Someday, September 14, 2015 will be noted as a pivotal failure for physics.

The author's fist book provided substantial details of LIGO's unjustified claims.

References has a link to a pdf having a detailed history of LIGO GW detections, from 2015 through the end of 2019. Each detection has the date of the earth tide event which LIGO reacted to. Sometimes LIGO triggers before that date, or on it, or after it, or sometimes more than once on one date. LIGO is not consistent.

Its title is: ligo-events4table.pdf

Note: This PDF is formatted for 8.5x11" page and has 2 pages. Compressing that content into a smaller page is impractical.

This is the same file referenced in the author's cosmology books.

Expensive projects like LIGO serve a small community of certain scientists.

There are no gravitational waves. This topic is only slightly relevant to gravity, the main topic of this book. Nothing more is required here.

10 Gravitational Lensing

Gravity cannot bend light but plasma in a star's atmosphere an electric field can bend light as confirmed with a number of observations.

The most famous test was the 1919 solar eclipse.
Sir Eddington observed a star during the eclipse, directly at the solar limb for the maximum diffraction through the bottom denser layer in the solar atmosphere.

Other observers noted other stars did not bend correctly. Einstein's prediction expects a deflection proportional to the distance from the star, with the maximum at the limb. No stars other than the one on the limb were affected as expected.

This rigged single experiment proved a changing medium of plasma, like in a star's atmosphere, can bend light.
Dr. Dowdye's experiments confirmed light bent as expected at different separations through the plasma for a changing density.
It did not confirm gravity bends light.

If it did, there should be distant stars appearing around every large star bending the light of the stars behind it, or perhaps around every large galaxy.

For those wishing a thorough explanation, there is a roughly 10 minute YouTube video explaining how light propagating through plasma will bend at a predictable and confirmed angle.

The video is titled "Can Stars BEND LIGHT? General Relativity and Gravity with Dr. Edward Dowdye!"

In simple terms, light bends by a change in the medium, just as observed by the Sun's position being altered relative to the horizon during a sunrise or sunset.
Tmost frequent use of gravitational lensing is avoiding the mistakes with red shifts.

The author's first book extensively explained red and blue shifts. Astronomers seem to remain ignorant of the fact galaxies have 2 red shift mechanisms and quasars have 2 red shift mechanisms and none of the 4 is the object's velocity.

When astronomers find a high red shift object where it should not be, then they explain it away by claiming a gravitational lens bent its light for an illusion with its observed awkward location. If an object is far from a visible object which could serve as the "lens" then dark matter is invoked, which compounds the confusion in dealing with red shifts.

When both galaxies and quasars can have either high or low red shifts, none of them a velocity, the combinations are endless.

11 Black Hole

A black hole does not exist but one is proposed for the M87 galaxy core.

This video explains a plasmoid and the object observed in M87, in April 2020. Its title:

Thornhill: Black Hole or Plasmoid? | Space News

Excerpt from Wikipedia for Plasmoid:

A plasmoid is a coherent structure of plasma and magnetic fields. Plasmoids have been proposed to explain natural phenomena such as ball lightning, magnetic bubbles in the magnetosphere, and objects in cometary tails, in the solar wind, in the solar atmosphere, and in the heliospheric current sheet. Plasmoids produced in the laboratory include field-reversed configurations, spheromaks, and in dense plasma focuses.
The word plasmoid was coined in 1956 by Winston H. Bostick to mean a "plasma-magnetic entity":
The plasma is emitted not as an amorphous blob, but in the form of a torus. We shall take the liberty of calling this toroidal structure a plasmoid, a word which means plasma-magnetic entity. The word plasmoid will be employed as a generic term for all plasma-magnetic entities.

(Excerpt end)

This fictitious entity was covered in both books.

A black hole has a simple purpose in popular cosmology.

Plasma and electromagnetic forces are ignored so synchrotron radiation is also ignored. As a result there is no available source for X-rays.

A black hole enables an impossible source of X-rays as a theoretical result of an extremely hot object. The solution was using a black hole which could supposedly hold an impossible mass within a point, or a sphere of zero diameter.

Excerpt from Wikipedia:

A black hole is a region of spacetime where gravity is so strong that nothing—no particles or even electromagnetic radiation such as light—can escape from it. The theory of general relativity predicts that a sufficiently compact mass can deform spacetime to form a black hole. The boundary of the region from which no escape is possible is called the event horizon. Although the event horizon has an enormous effect on the fate and circumstances of an object crossing it, according to general relativity it has no locally detectable features.

In many ways, a black hole acts like an ideal black body, as it reflects no light. Moreover, quantum field theory in curved spacetime predicts that event horizons emit Hawking radiation, with the same spectrum as a black body of a temperature inversely proportional to its mass. This temperature is on the order of billionths of a kelvin for black holes of stellar mass, making it essentially impossible to observe.

(Excerpt end)

Observation:

All is "according to relativity" which involves only the moving observer's reference frame. No other observer observes curvature, so no other observer, like one on or near Earth can observe the curvature at a black hole.

The impossible claims of a black hole were detailed in the author's first book. Many others have their own way to justify the inevitable conclusion there are no black holes.

Even Albert Einstein knew a black hole is impossible, with his famous quote:

"Black holes are where God divided by zero."

A black hole is fictional.

All objects claimed to be a black hole are sources of synchrotron radiation indicating a plasmoid and not a black hole with its impossible accretion disk.

The infamous fabricated image in April 2020 of an object in M87 galaxy was not a black hole. It was a plasmoid, covered in a previous section.

A plasmoid has a torus shape. The fabricated image had to have a ring because an accretion disk was expected,

No black hole has ever been detected because they don't exist.

The first book detailed the false claims of LIGO which never detected a non-existent gravitational wave, though LIGO made false claims of mergers with black holes. LIGO never detected an astrophysical event of any kind (all events were actually terrestrial), and certainly not with a black hole or neutron star.

A story in April 2020 was titled:
Astronomers saw a star dancing around a black hole. And it proves Einstein's theory was right

Observation:

 For many years astronomers have been seeking a star orbiting around the claimed black hole at the center of our Milky Way. If its orbit is a valid Kepler ellipse then they seek to apply Kepler's 3rd law to calculate the MBH mass. This was done initially with the star S1 several years ago to derive the unjustified number of billions of solar masses. Unfortunately, someone noticed S1 was not a valid ellipse; instead of the claimed period of 16 years, its claimed radius required many thousands of years. The invalid mass persists but astronomers seek another star to try again for a new mass calculation.
Obviously S2 orbit is invalid as well because it is not closed, as stated in the article.
Because astronomers have been watching for a few decades, so perhaps this time, they gave up and came up with this "rosette" orbit to bring up Einstein - which is obviously effective to get your story published.

From Wikipedia, not the story:

S2, also known as S0–2, is a star that is located close to the radio source Sagittarius A*, orbiting it with an orbital period of 16.0518 years, a semi-major axis of about 970 au.

(Excerpt end)

Observation:

S2 = 970 au, 16.052 y

Jupiter= 5.203au, 11.86 y
Saturn= 9.54au, 29.46y

With Kepler's 3rd law:

S2 = 970 au takes about 30,200yr

S2= 16.0518 yr is for about 6.5au

the observed orbit parameters for S2 are not a valid Kepler ellipse.

These news stories never reveal the importance of this search for a star like S2 and the requirement for a valid ellipse.

With a valid ellipse, astronomers can use Newton's change to Kepler's 3rd law to calculate the number of solar masses at the core with a known mass in an orbiting star. The mass for S2 will be a guess. With electromagnetic forces in play this calculation might be debated for many reasons.

Currently, every black hole claimed to be in every galaxy is assigned a value of solar masses loosely based on the estimated number of stars in that galaxy. That assumption has no basis.

If the Milky Way core, which is claimed to have a SMBH for the roughly 300 billion stars in the galaxy, is monitored to seek an orbiting star. When finally found, this will be a significant milestone for black hole advocates.

Until that star is found, there is no black hole in any galaxy with a justification for its claimed number of solar masses.

This is a basic problem for cosmologists of no evidence or no justification for any claims about a SMBH in every galaxy. Cosmologists hope this finding and calculation can salvage all their claims spanning many years. Unfortunately for them, an orbit with a larger radius for its measurement requires a longer period like in thousands of years (like S2 needs).

One might expect rules to be broken about the orbit parameters to achieve the crucial calculation.

Every scientific claim requires evidence.

Currently, there is no evidence for any value of a black hole's mass because a correct measurement has never been done.

If a measurement ever succeeds for a star in motion around a possible "hole" in space, whatever is in that "hole" it is not a black hole defined by relativity, because a black hole, with mass residing in a singularity, is impossible by physics.

12 Neutron Star

Wikipedia excerpt:

A neutron star is the collapsed core of a massive supergiant star, which had a total mass of between 10 and 25 solar masses, possibly more if the star was especially metal-rich. Neutron stars are the smallest and densest stellar objects, excluding black holes and hypothetical white holes, quark stars, and strange stars. Neutron stars have a radius on the order of 10 kilometres (6.2 mi) and a mass of about 1.4 solar masses. They result from the supernova explosion of a massive star, combined with gravitational collapse, that compresses the core past white dwarf star density to that of atomic nuclei. The origins of the strong magnetic field are as yet unclear.

(Excerpt end)

Observation:
This entire explanation is invalid. There is no evidence many neutrons with no protons will not decay just like a solitary neutron decays in a few minutes. The description admits the magnetic field cannot be explained.

There is no evidence of a force capable of creating and maintaining a neutron star, whether gravity, the strong force in a normal nucleus or some other, unidentified force.

Thunderbolts Project has a YouTube video clearly explaining a neutron star to match observations, titled:

The Invention of the Neutron Star | Space News

Observation:

Gravity alone cannot explain a supposed neutron star due to the physically impossible conditions being proposed. With no evidence, a neutron star is just claimed to exist.

They are required in cosmology because a plasmoid is never considered as a source of the observed synchrotron radiation, though a plasmoid is the only viable explanation.

Section 8 detailed the false claims of LIGO which never detected a non-existent gravitational wave, though LIGO made false claims of mergers with neutron stars.

Neutron stars do not exist.

They are not important to gravity.

13 Gravitational Precession

Definition from Wikipedia:

Precession is a change in the orientation of the rotational axis of a rotating body. In an appropriate reference frame it can be defined as a change in the first Euler angle, whereas the third Euler angle defines the rotation itself. In other words, if the axis of rotation of a body is itself rotating about a second axis, that body is said to be precessing about the second axis. A motion in which the second Euler angle changes is called nutation. In physics, there are two types of precession: torque-free and torque-induced.
In astronomy, precession refers to any of several slow changes in an astronomical body's rotational or orbital parameters.

An important example is the steady change in the orientation of the axis of rotation of the Earth, known as the precession of the equinoxes.
(Excerpt end)

A number of scholars have explained the precession of Mercury's orbit by using Newton, not Einstein. One example is the 2015 paper titled:

Explanation of the Perihelion Motion of Mercury in Terms of a Velocity-Dependent Correction to Newton's Law of Gravitation

14 Dark Matter

Dark matter is the proposed solution when gravity is assumed to be the cause of a behavior but not enough visible matter can be found.

Rather than seeking an unseen force, an explanation using physics is abandoned. Instead dark matter is proposed with the hope some day this extemporaneous excuse is offered.
This will be explained below with references but a spiral galaxy rotates by the galactic magnetic field and there is no dark matter.

Nearly all of the universe is plasma or matter with an electrical charge. Its short list is electrons, protons, and ions.

Plasma has unique behaviors. Plasma is not a "gas" but it naturally forms filaments. The flow of charges creates a magnetic field which constricts the flow, into a filament.

Cosmologists often ignore plasma behaviors so observed filaments sometimes require dark matter as the "cause" of its structure, because plasma behaviors are ignored resulting in no valid explanation so the excuse for finding no real cause is called dark matter.
In most cases where dark matter is proposed there is an unseen magnetic field affecting plasma.

14.1 Dark Matter Beginnings

Dark matter arose in the 1930's when insufficient data resulted in the conclusion the observational data could not match expectations.

Excerpt from Wikipedia about Fritz Zwicky:

While examining the Coma galaxy cluster in 1933, Zwicky was the first to use the virial theorem to discover the existence of a gravitational anomaly, which he termed dunkle (kalt) Materie 'dark matter'. The gravitational anomaly surfaced due to the excessive rotational velocity of luminous matter compared to the calculated gravitational attraction within the cluster. He calculated the gravitational mass of the galaxies within the cluster from the observed rotational velocities and obtained a value at least 400 times greater than expected from their luminosity. The same calculation today shows a smaller factor, based on greater values for the mass of luminous material; but it is still clear that the great majority of matter was correctly inferred to be dark.

(Excerpt end)

Observation:

With modern numbers, the factor dropped from 400 to "great majority"

In 1933, Zwicky's explanation for an anomaly was believed, though lacking accurate data to prove his conclusion and with no evidence that other causes might have been either sought or missed.

13.2 Dark matter replaced by galactic magnetic field

Andromeda Galaxy or M31 was found to have an unexpected rotation curve.

The stars in the disk were assumed to have Keplerian orbits as if billions of stars move just like the planets around our Sun.

When that expectation failed by observation, dark matter was offered as the ad hoc explanation, rather than revising the expectation.

Image and caption from Wikipedia:

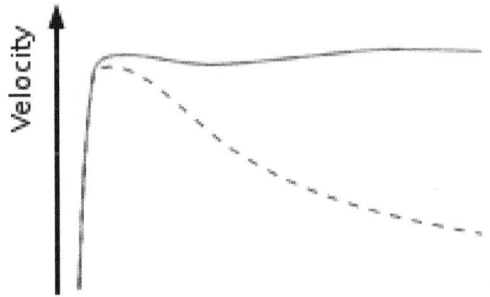

Rotation curve of a typical spiral galaxy: predicted (A) and observed (B).

The discrepancy between the curves is attributed to dark matter.

Observation:

The critical word is predicted. The discrepancy is caused by a wrong model, which was based on the unjustified assumption that billions of stars will rotate in a disk spanning thousands of light-years just like 8 planets around our Sun. They expected Keplerian orbits but the data failed to match the prediction. Their response was dark matter (which remains undefined), rather than revising their model to match observations, as is commonly done in physics.

Several studies reached the same conclusion about spiral galaxies: the galactic magnetic field drives the disk rotation. The M31 magnetic field has been measured.

The Journal Progress in Physics in April 2018 had the paper by Donald Scott titled:

Birkelund Currents and Dark Matter

Link is in References.

Figure 6 in the paper shows how well the predictions for a spiral galaxy rotation closely match observations.

Comparison of Predicted and Actual Stellar Velocity Profiles

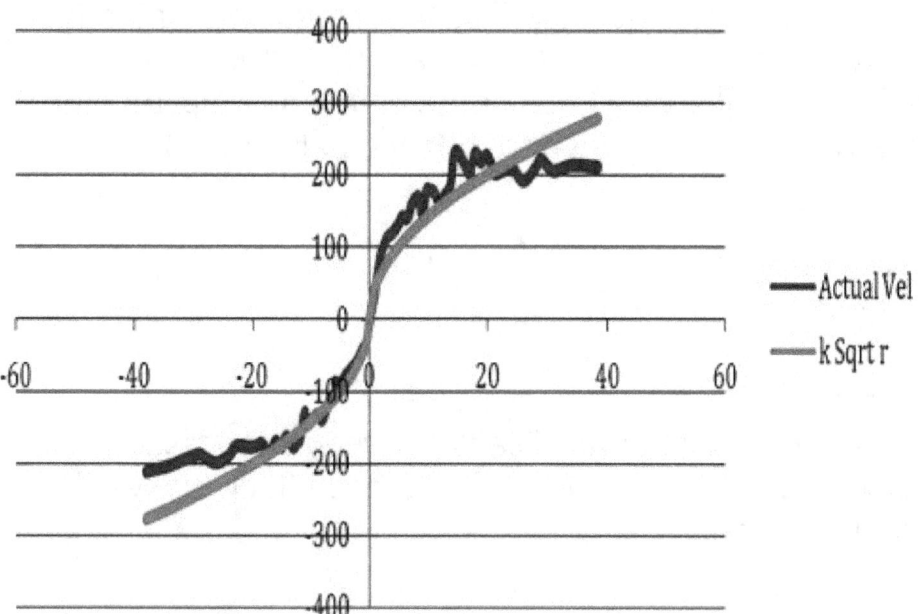

Fig. 6: Comparison of the example galaxy's measured velocity profile with the Bessel function model's Sqrt r profile.

The paper provides more details.

For those needing more references about a spiral galaxy rotation curve, scientists in Spain published the paper titled:

MAGNETIC FIELDS AND THE OUTER ROTATION CURVE OF M31

Excerpt:

It is certainly a challenge as the standard dark matter halo models, in particular the universal NFW profiles, do not account for this dynamical unexpected behavior.
Our conclusion is that a significant improvement of the fit in the outer part is obtained when magnetic effects are considered. The best-fit solution requires an amplitude of ~4 µG with a weak radial dependence between 10 and 38 kpc.

(Excerpt end)

Observation:
Even the "dark matter halo" fails to account for "this unexpected behavior" but the "magnetic effect" is the "best-fit solution."

14.3 IC342 Spiral Galaxy

A study revealed the importance of the magnetic fields in the spiral arms

The story:

Twisted magnetic field in galaxy IC 342

An interesting conclusion after a study of IC342, a large obscured, nearby spiral galaxy:

Excerpt:

Magnetic fields exist everywhere in the Universe, but what role do they play in the evolution of cosmic objects? In their detailed data of the nearby galaxy IC 342 from observations with two of the world's largest radio telescopes, Astronomers at the Max Planck Institute for Radio Astronomy (MPIfR) have discovered a magnetic field aligned along the optical spiral arms. "Our observations can help to discover answers to the question of how galaxies evolve and develop further," says project leader Rainer Beck.
"Spiral arms can hardly be formed by gravitational forces alone," continues Rainer Beck. "This new IC 342 image indicates that magnetic fields also play an important role in forming spiral arms."

(Excerpt end)

Observation:

Dark matter is proposed as the reason for structure in a spiral galaxy.

Excerpt from NOAO:

Spiral Galaxy IC342 is located roughly 11 million light-years from Earth. Its face-on appearance in the sky—as opposed to our tilted and edge-on views of many other nearby galaxies, such as the large spiral galaxy Andromeda (M31)—makes IC342 a prime target for studies of star formation and astrochemistry.

(Excerpt end)

Excerpt from another source:

[This] galaxy is obscured by the dust in our galaxy by nearly three full magnitudes. If it weren't for this dimming effect, IC342 would be one of the finest and brightest face-on spirals in the sky and easily visible to the unaided eye in dark sky. That would make it by far the most distant object you could see with the unaided eye.

(Excerpt end)

Observation:
The conclusion about the important role of magnetic fields presents a problem for cosmologists because those fields are usually not even looked for.

Currently the rotation curve of a spiral galaxy is assumed to be driven by the mass distribution in the disk.
This scientist claims gravity alone does not drive the spiral arm formation.

This is an important conclusion because when the mass distribution does not result in the correct rotation curve, dark matter is proposed to "fix" the "required" mass distribution.

-

Studies like this should lead to the removal of dark matter.

14.4 Rings and arcs

Rings and arcs are often around giant elliptical galaxies, but not other galaxy types.

The following objects represent some of those images. These objects are in no particular order.

Hoag's Object is a known ring galaxy.

From Wikipedia:

In the initial announcement of his discovery, Hoag proposed the hypothesis that the visible ring was a product of gravitational lensing. This idea was later discarded because the nucleus and the ring have the same redshift, and because more advanced telescopes revealed the knotty structure of the ring, something that would not be visible if the ring were the product of gravitational lensing.

(Excerpt end)

The criteria about knots is not applied consistently because many other rings show knots but are still claimed to be a distortion by a lens.

There are other ring galaxies

Cartwheel Galaxy
Impressive ring with large radius, with faint spokes

Here is another ring galaxy:

AM 0644-741 or Lindsay-Shapley Ring

It looks like Hoag's Object but the core is left of center

The following objects are rings and arcs claimed to be distortions.

They are in no particular order. All have links in References.

RXJ1131-1231 - more complex Hoag's object

SDSS J0146-0929 - multiple arcs as parts of a circle

Abell 2261 - arc at 9 o'clock; some documents mention knots in Abell 2261.

Abell 383 - long arc at 4 o'clock ending with a knot; beyond that at 5 o'clock is elliptical with distorted circle.

Abell S1063 - several arcs around the elliptical

Abell 1413 - arc at 10 o'clock

Abell 2218 - multiple arcs around the large elliptical at left; filaments around elliptical near top right

Abell 1689 - faint arcs at 2 and 4 o'clock

Abell 2261 - arc at 9 o'clock from large elliptical

Abell 2390 - in X-ray both images show arcs

Abell 2667 large arc to left of large elliptical, the bottom of the arc has a knot

Abell 370 - large arc at 2 o'clock to central elliptical; the bottom of the arc has a knot

SDSS J1004+4112 - the central bright object is claimed to be a lensed quasar but it looks like an elliptical with a long jet at about 10 o'clock; there is an arc with knots at 4 o'clock with another small arc above it; there is also an arc at 2 o'clock that crosses a distant galaxy - quite the coincidence!

When zooming into an image there is another elliptical some distance away from the main one at about 9 o'clock. Oddly this one has an arc with knots at 10 o'clock. Near the right edge is either: two galaxies are merging or one is splitting. Below that is two spirals with long tails. This is a very interesting cluster.

SDP.81 - almost a complete circle with a knot inside the circle

LRG 3-757 - the horse shoe Einstein ring

RCS2 032727-132623 - a very large arc around the central elliptical; odd filaments at 4 and 12 o'clock.

Abell 2261 - arc at 9 o'clock to the BCG

J1531+3414 - cluster with much of interest
double ellipticals at center of very large diameter ring

ZwCl0024+1652 (CL0024+17 for short) is a galaxy cluster
with a ghostly ring of dark matter
Some images show a wide diffuse ring around
the ellipticals at center

Another image shows many arcs around the central
ellipticals

SDSS J103842.59+484917.7 - Cheshire Cat

Story has images in different wavelengths

Summary of collection of images:

Many of these distant galaxy clusters appear to exhibit much electrical activity as arcs.
The fine rings and arcs appear part of a circular plasma filament around a large elliptical galaxy where the visible portion is in arc mode and the rest is in dark mode.

There is a pattern for which galaxy type is at the center of observed arcs.

If an arc is claimed is claimed to be from a gravitational lens effect than every massive galaxy should be capable, but no spiral galaxies have arcs.

M31 has a trillion stars but no arcs are observed around this massive object.

There must be more of these "distortions" than I could find, but this was a worthwhile sample.

Some supposed collisions of galaxies are in sensational stories from astronomers.

Collisions of galaxies have no basis. As noted in the first book, no galaxy ever had its proper 3-dimensional motion measured. To propose a collision there must be evidence for motion implying either a collision or a process of fissioning, the opposite interaction.

Gravity cannot be proposed as causing a collision of galaxies when lacking any justification for an assumed direction of motion.

14.5 Laniakeia

Excerpt from Wikipedia:

The Local Supercluster, Local SCI, or Laniakea Supercluster (Laniakea, Hawaiian for open skies or immense heaven), is the galaxy supercluster that is home to the Milky Way and approximately 100,000 other nearby galaxies. It was defined in September 2014, when a group of astronomers including R. Brent Tully of the University of Hawaii, Hélène Courtois of the University of Lyon, Yehuda Hoffman of the Hebrew University of Jerusalem, and Daniel Pomarède of CEA Université Paris-Saclay published a new way of defining superclusters according to the relative velocities of galaxies. The new definition of the local supercluster subsumes the prior defined local supercluster, the Virgo Supercluster, as an appendage.
Follow-up studies suggest that the Local Supercluster is not gravitationally bound; it will disperse rather than continue to maintain itself as an overdensity relative to surrounding areas.

(Excerpt end)

Observation:

"Local Supercluster is not gravitationally bound" so the only other force at play here is the magnetic force which can affect plasma and objects having electric fields.

However they reached this conclusion though lacking velocity data for all objects, gravity is being excluded.

Dark matter does not exist so it is not important to gravity.

I noticed other attempts at explaining gravity include dark matter, as if it were relevant. Such attempts should be ignored because of their initial, wrong foundation.

14.5 Other matters

Here is an author's comment to the reader before the book's conclusion.

Someone could ask the legitimate question:

Why publish a paperback, not an academic paper which could be read and reviewed by the scientific community?

Gravity is a fundamental force.

The answer is simply the book has a slightly higher probability of being noticed.

A study published in 2010 concluded the M31 galactic magnetic field explained its rotation curve, not dark matter. A study published in 2015 concluded the IC 342 spiral arms have structure from magnetic fields not gravity explained their formation [not dark matter].

The pair of studies removed the justification for dark matter.

The 2019 Nobel Prize in Physics went to someone working many years on a fruitless effort to explain dark matter. The award followed the studies. Dark matter is a blatant mistake, caused by cosmologists ignoring electromagnetic forces. Dark matter does not exist. Anyone working on it is just wasting their time and resources.

On November 10, 2019, I predicted LIGO detections using specific dates, by their terrestrial source. I notified NSF of this demonstration of LIGO's mistake. My contact was acknowledged but there was no change or public revelation of LIGO's false claims. LIGO had received a Nobel Prize in 2017 for its false claims. There has been no retraction of the award. The story of LIGO was told in the first book.

This book changed its emphasis during its writing. Initially the relativity aspect was important, almost as important as the force of gravity mechanism. When the initial book was conceived, mass defect was a detail to complete later, after finishing the rest.
The mass defect became an important observation requiring a detailed explanation.

The result of a changing emphasis is the narrative through the book is probably different than if the book were just restarted to weave the many pieces in a slightly different manner.

Modern sciences lack the discipline to maintain their integrity. Defective theories and dogma persist. I noted there is no published margin of error for an atomic mass value though the calculation for a mass defect absolutely requires a known margin of error for a result to justify its high precision. This lack of clarity is compounded by the omission of any indication of when an atom's measured mass value is actually the result of the element's isotope ratio.

I am retired with no academic or corporate affiliations.
My simple goal is improving several sciences, initially by
posting and later by writing several books.
Much data are out there, but apparently I integrate them
differently than others.
Books can be lost, but also can be archived.

. This book mixes many topics so a coherent narrative is
difficult. Discovering public data of the elements is
sometimes wrong is disappointing.

I hope you enjoyed this book, and found it was worth your
time. I am probably not correct all the time, but I expect my
conclusions are suitable for consideration, based on the
current data.

Sometimes, time is more precious than money, depending
on how you spend the time.

This author began this book by expanding on the concept
of his simple particle model, which was posted to a
facebook group on September 20, 2020. That concept was
the basis for pursuing a mechanism for gravity.

The mass defect is a known anomaly for measuring the
mass of an atom. It needed an explanation, requiring
research and analysis. All that analysis did not exist at the
time of that post.

The presentation of analysis might not be up to
professional standards, when confronted with wrong
atomic mass values. This book cannot explicitly describe
the mass defect mechanism, because the author lacks the
expertise and resources for that task.

If this book were titled "redefining particle physics" then no one would take the risk at reading a book making such a false claim.

Mass defect requires a verified explanation. The goal of this book is getting those competent in their sciences to do a better job, by pointing out new paths which should be more productive.

The 4 books cover many topics in physics and more.

For example, the first book identified the mistake of treating a red shift as a velocity. This should have been fixed in 1936 when Hubble noted the context of the anomaly for galaxies beyond our Local Group, but it was not fixed at the crucial time of starting modern cosmology so the big bang fiction persists.

15 Final Conclusion

Relativity has several notable mistakes:

a) space-time is not an acceptable replacement for gravity,

b) gravity is instantaneous, and does not have a velocity at c or the same as light,
c) gravitational waves don't exist,

d) gravitational lensing cannot occur,
   light bends in plasma,
d) gravitational precession does not require relativity,
e) black holes don't exist,
f) matter does not have a velocity limit at c.

The force of gravity defined by Isaac Newton remains valid for physics, astrophysics, and cosmology.

Newton admitted he could not explain how his force worked over a distance, though the force has been verified numerous times.

This omission enabled Einstein to propose a behavior like gravity but with no force. This concept is called space-time, which is limited to only a special observer.

The author proposed a mechanism for Newton's fundamental force of gravity, using physics, not a coordinate system.

Newton did not understand electric fields which were explained by James Clerk Maxwell and others, many years later.

Relativity and its interpretation of gravity as defined by Albert Einstein are not valid for physics, astrophysics, and cosmology.

In this book, gravity has been redefined to use Newton's force of gravity, which has been thoroughly verified.

This book offers 5 contributions to physics from the author:

1) new mass values for the proton and electron,
2) a new behavior for a neutron in an atom,
 which is required to explain atomic mass defect.
3) a mechanism for Newton's force of gravity,
4) a different mechanism for particle pair production,

5) a new version of Kepler's third law,

Item (5) could be considered trivial, but it should be correct.

Consequences:

Item (1) is probably controversial.

The bottom line for this decision making such a fundamental change in particle physics is whether all agree the expected mass of the hydrogen atom, ^1H, must be the sum of a proton and electron.

The basis for this change involves only 2 values:
1) Using ^1H not ^{12}C for the benchmark atom.

The presence of nuclear binding energy in ^{12}C is probably why the particle masses are wrong for ^1H.

The current mass value for ^{1}H was not changed, assuming it was properly verified.

2) Using the ratio between a proton's mass an electron's mass.

This ratio must have substantial experimental evidence to justify its current value. To get the recommended new masses for the two particles this value was not changed.

The only change is in the benchmark atom from ^{12}C to ^{1}H, which is also called protium.
The immediate impact of particle mass changes is the ^{1}H atomic mass exactly matches the sum of its 2 components. With the component masses derived from ^{12}C, the sum is too high for ^{1}H. This difference is called an atomic mass defect.

This change fixes the problem of nuclear binding energy loss in a nucleus having only one proton, so there is nothing to bind. Apparently, physicists do not see this crucial deviation as a problem, because the wrong particle masses have persisted.

This is a fundamental problem with the current mass values for a proton and electron.

This also means all atoms will have a slightly lower mass defect. The value for ^{1}H becomes zero.

Item (2) is certainly controversial.

The neutron is just a particle in the Standard Model having no defined behaviors beyond its mass and no charge.

Now, the neutron is a dynamic partner within the atomic nucleus and its reactivity to external masses.

I have no means to test
a) my mechanism for gravity,
b) neutron's apparent mass changing, while within the nucleus, when being measured outside the atom.

However, my mechanism is simple.

Item (3) is probably controversial because the current mechanism remains unchallenged.

This recommendation involves a simple mechanism involving an electron flipping its polarity when the atom absorbs a high energy gamma ray.

I have no means to test my mechanism for pair production.

However, my mechanism is simple.

The current mechanism proposes 2 particles are created. My mechanism creates neither.

Item (4) is probably controversial because the current mechanism, apparently proposed by Einstein, remains unchallenged.

The mass defect can be explained by neutrons changing their observed "mass" when locked in a tight bond with one or more protons. A mass defect is noted only in the neutrons.

It is observed in the neutron formation and in deuterium, which is only a proton and a neutron. It is also observed in many atomic nuclei. In atoms, the difference can be an increase or decrease, depending on the particular element.

The change in particle mass values arose from using the published atomic mass value for protium because this atom contains only the two fundamental particles. Their respective masses can be calculated from a measurement of their sum, the protium atom.
This is very straightforward, rather than using the carbon-12 atom as now, which has 6 protons, 6 neutrons, and 6 electrons.
The mass defect analysis uses a consistent presentation of atomic mass values in Wikipedia, so there should be no question of the source of the values.

There were several elements having a published atomic mass value which should explicitly state the value is from an isotope ratio and is not a true measured value.

As a result astrophysics students have calculation exercises for the binding energy based on the incorrect mass for the element.

Students should learn with correct data and when learning mass defect, they should learn the correct way and be told of misleading public atomic mass values of the elements.

Each of the author's earlier 3 books also offered unique contributions from the author, which could help the progress of several sciences.

16 References

The references in this book are available as clickable links from a page in the author's web site.

1. Start web browser.

2. Go to this site: www.cosmologyview.com

3. Make sure the browser is on the correct home page:

Cosmology Views

4. Scroll to near the middle.

5. Select: **Books by the author**

This page presents information for each book.

Locate the columns for this book.

6. Locate: **Redefining Gravity**

7. Below it, locate the date of this book's edition: 10/13/2020

8. Select: **References** after the correct date.

The selected page will list the references in the book by page number, with a link to that reference.